Inverse and Ill-Posed Problems Series

Composite Type Equations and Inverse Problems

Also available in the Inverse and Ill-Posed Problems Series:

Inverse Problems of Electromagnetic Geophysical Fields
P.S. Martyshko

Inverse Problems of Vibrational Spectroscopy
A.G. Yagola, I.V. Kochikov, G.M. Kuramshina and Yu.A. Pentin

Elements of the Theory of Inverse Problems
A.M. Denisov

Volterra Equations and Inverse Problems
A.L. Bughgeim

Small Parameter Method in Multidimensional Inverse Problems
A.S. Barashkov

Regularization, Uniqueness and Existence of Volterra Equations of the First Kind
A. Asanov

Methods for Solution of Nonlinear Operator Equations
V.P. Tanana

Inverse and Ill-Posed Sources Problems
Yu.E. Anikonov, B.A. Bubnov and G.N. Erokhin

Methods for Solving Operator Equations
V.P. Tanana

Nonclassical and Inverse Problems for Pseudoparabolic Equations
A. Asanov and E.R. Atamanov

Formulas in Inverse and Ill-Posed Problems
Yu.E. Anikonov

Inverse Logarithmic Potential Problem
V.G. Cherednichenko

Multidimensional Inverse and Ill-Posed Problems for Differential Equations
Yu.E. Anikonov

Ill-Posed Problems with A Priori Information
V.V. Vasin and A.L. Ageev

Integral Geometry of Tensor Fields
V.A. Sharafutdinov

Inverse Problems for Maxwell's Equations
V.G. Romanov and S.I. Kabanikhin

INVERSE AND ILL-POSED PROBLEMS SERIES

Composite Type Equations and Inverse Problems

A.I. Kozhanov

///VSP///
Utrecht, The Netherlands, 1999

VSP BV
P.O. Box 346
3700 AH Zeist
The Netherlands

Tel: +31 30 692 5790
Fax: +31 30 693 2081
E-mail: vsppub@compuserve.com
Home Page: http://www.vsppub.com

First published in 1999

ISBN 90-6764-305-X

Printed in The Netherlands by Ridderprint bv, Ridderkerk.

Contents

Preface

The results stated in this monograph do not give a full representation of the development of the theory of composite type equations and its applications. We do not discuss the results on nonlinear composite type equations. Also, we do not set forth the results on qualitative properties of solutions (increment of the smoothness, behavior depending on parameters, and so on). We do not state the problems of setting and solvability of boundary-value problems for operators whose higher parts represent the superposition of hyperbolic and hyperbolic-parabolic operators. In some cases, these results have been stated but they were not obtained by the author; in other cases, the results do not accord with the monograph conception or else have not been obtained.

However, the author is sure that all the questions and problems enumerated above will be set forth in further papers and monographs.

Introduction

Nowadays, in the modern mathematical literature, the theory of boundary-value problems for equations of composite type of the third order with two independent variables is stated rather satisfactorily (in this case, the methods of complex function theory work effectively). However, in the case of many independent variables, in most cases, these results are set forth. These results correspond to the case of equations with a selected variable in special domains. Our monograph is an attempt to complete this presentation - we have tried to set forth the theory of well-posed boundary-value problems for general equations of composite type of the third order in the domain of arbitrary dimension and geometry.

In the monograph we mainly state the results of the author.The structure of the book is as follows. In Chapter I we consider the results on solvability of boundary-value problems and on the properties of equations of the composite type of the third order. Chapter II is devoted to the problems of applications of the theory developed in Chapter I and to the proof of solvability of certain inverse problems for the second order equations. Also, here we have considered some new boundary-value problems for nonclassic equations of the second order and new boundary-value problems for a strongly damped wave equation.

For research of boundary-value problems, we use mainly the well-known schemes based on obtaining *a priori* estimates. As a rule, we construct approximate solutions on the basis of regularization of either the initial problem, or the problem well investigated in previous works, or the problem investigated before in our monograph.

Notations in the monograph are as usual. The numeration of theorems, lemmas, and formulas is triple. It shows the number of the chapter, paragraph, and the running order of the theorem, lemma or formula. The *a priori* estimates which arise in proofs are designated by the letters C, N, K and so on with indices. As a rule, these constants will differ in different paragraphs. In this case, the phrase "the constant C depends on the function f" will denote that this constant depends on the norm of the function f in the space mentioned in the formulation of the statement in which proof this constant had arisen.

Chapter 1.

Composite type equations as the original mathematical object

1.1. EQUATIONS OF THE COMPOSITE TYPE IN MATHEMATICS AND MATHEMATICAL SIMULATION

The interest which has recently arisen in equations of odd order, in particular, of the third order, may be explained from two points of view. On the one hand, these are the requirements of mechanics, hydro- and gas dynamics and other applied sciences. On the other hand, we have the development of mathematics in itself. So, in mechanics and physics, the well known equations are the long wave equations - the Korteweg-de Vries equations and the alternative Korteweg-de Vries equations respectively

$$u_t + u_{xxx} + uu_x = f,$$

$$u_t - u_{xxt} + uu_x = f.$$

In the theory of material elasticity with the memory and in visco-elasticity there arise the equations

$$u_{tt} - \frac{\partial}{\partial x}F(u_x) - \frac{\partial}{\partial t}\frac{\partial}{\partial x}F_0(u_x) = f$$

(see, Maslov and Mosolov, 1990). The analogous equations arise when modelling the processes of liquid transfer (see Nakhushev, 1995). When we investigate hydrodynamics of non-Newtonian liquids, we add derivatives of the third and higher orders (Oskolkov, 1975, 1980).

The equations of composite type, as independent mathematical objects arose first in the works of Hadamard (1933, 1936). The researches of J. Hadamard were continued by O. Sjöstrand, L. Cattabriga, L. Wolfersdorf and others. Various aspects of the theory of composite type equations were considered in monographs of Salakhitdinov (1974), Dzhuraev (1979). In all these works the equations in question are investigated mainly in the plane and with the model operators in the senior part. Certain classes of equations of the composite type in multidimensional domains were investigated by Grisvard (1967), Dubinskii (1973), Berkhin (1974), Vragov (1978), Terekhov (1980), Meredov (1981), Daynyak and Korzyuk (1987) and in some other works. However, in all these works there were considered either equations with a selected variable, or generalized solutions were obtained.

In recent years, special equations of the composite type have been actively investigated. These are the pseudoparabolic equations and pseudohyperbolic equations, respectively,

$$u_t - Au_t - Bu = f$$

and

$$u_{tt} - Au_t - Bu = f,$$

where A and B are linear or nonlinear operators of even order (pseudoparabolic and pseudohyperbolic equations which are often called equations of the Sobolev type).

The equations of composite type are connected with so-called nonclassical equations. These are degenerate equations of second and higher orders. Keldysh (1951) was the first to show that the settings of boundary-value problems for degenerate equations are essentially different from the settings of boundary-value problems for nondegenerate (classical) equations. In the works of Fichera (1960), Oleinik and Radkevich (1969), Karatoprakliev (1977), Vragov (1977), for the various classes of degenerate equations and equations of the composite type, there were suggested the settings of boundary-value problems, where the supports of boundary data were determined by coefficients of the equation. Analogous results have been obtained by the author for the composite type equations and they are stated below.

In conclusion, we may say that the theory of composite type equations as part of the general theory of partial differential equations is actively developing. We may also note that the equations of composite type are actively applied for research of second order equations, both classical and nonclassical (we shall consider them in more detail below).

Thus, we may say that the development of the theory of composite type equations is necessary both from the point of view of mathematical simulation theory and the general theory of partial differential equations.

1.2. CANONICAL TYPES OF THIRD ORDER EQUATIONS

The possibility of reducing the equation to a certain canonical form plays an important role in the theory of partial differential equations. This allows us both to determine the canonical types of boundary-value problems and to develop specific methods to solve this type of equation.

As is well known, in the case of two independent variables, partial differential equations, in each point where they do not degenerate into a first order equation or into an ordinary differential equation with a parameter, may be reduced to one of the following canonical types: elliptic, parabolic, or hyperbolic. As is also known, such membership of initial equations to one of the types may be described in the domain of definition in terms of coefficients of the equation. Equations with many partial derivatives may not always be elliptic, parabolic, or hyperbolic (at the point where they do not degenerate into an equation of the first order or into an equation with fewer variables). The equation may turn out to be ultraparabolic, ultrahyperbolic, or, in general, nonclassical. However, for equations with many variables, we may guarantee that they will be elliptic, hyperbolic, or parabolic type in a certain domain in terms of the coefficients of these equations.

In the case of third order equations, the situation is essentially more complicated.

Let us consider the equation with two independent variables and with constant coefficients

$$Au_{xxx} + Bu_{xxy} + Cu_{xyy} + Du_{yyy} + au_{xx} + bu_{xy} + cu_{yy}$$

$$+mu_x + nu_y + du + f(x, y) = 0 \qquad (1.2.1)$$

Dzhuraev and Popelek (1991) have shown that with the help of a nonsingular transformation, this equation may be reduced to one of the following types

$$\frac{\partial}{\partial\xi}\Big(u_{\xi\xi}+u_{\eta\eta}\Big)\pm\Big(u_{\xi\xi}+u_{\eta\eta}\Big)+\alpha u_\xi+\beta u_\eta+\gamma u+g(\xi,\eta)=0, \tag{1.2.2}$$

$$\frac{\partial}{\partial\xi}\Big(u_{\xi\xi}-u_{\eta\eta}\Big)\pm\Big(u_{\xi\xi}-u_{\eta\eta}\Big)+\alpha u_\xi+\beta u_\eta+\gamma u+g(\xi,\eta)=0, \tag{1.2.3}$$

$$u_{\xi\xi\eta}\pm u_{\xi\eta\eta}\pm u_{\xi\eta}+\alpha u_\xi+\beta u_\eta+\gamma u+g(\xi,\eta)=0, \tag{1.2.4}$$

$$u_{\xi\xi\eta}\pm u_{\eta\eta}+\alpha u_\xi+\beta u_\eta+\gamma u+g(\xi,\eta)=0, \tag{1.2.5}$$

$$u_{\xi\xi\eta}\pm u_{\xi\eta}+\alpha u_\xi+\beta u_\eta+\gamma u+g(\xi,\eta)=0, \tag{1.2.6}$$

$$u_{\xi\xi\eta}\pm u_{\xi\xi}+\alpha u_\xi+\beta u_\eta+\gamma u+g(\xi,\eta)=0, \tag{1.2.7}$$

$$u_{\xi\xi\xi}\pm u_{\eta\eta}+\alpha u_\xi+\beta u_\eta+\gamma u+g(\xi,\eta)=0, \tag{1.2.8}$$

$$u_{\xi\xi\xi}\pm u_{\xi\eta}+\alpha u_\xi+\beta u_\eta+\gamma u+g(\xi,\eta)=0, \tag{1.2.9}$$

$$u_{\xi\xi\xi}\pm u_{\eta\eta}+\alpha u_{\xi\eta}+\beta u_\xi+\gamma u_\eta+\delta u+g(\xi,\eta)=0, \tag{1.2.10}$$

$$u_{\xi\xi\xi}\pm u_\eta+\alpha u_{\xi\xi}+\beta u+g(\xi,\eta)=0, \tag{1.2.11}$$

where α,β,γ, and δ are certain numbers dependent on the coefficients of equation (1.2.1).

Analyzing these equations, we may see that equations (1.2.3) and (1.2.4) are hyperbolic equations of the third order. The remaining equations (1.2.2), (1.2.5)–(1.2.11) may be written in the form

$$lAu+Bu+g(\xi,\eta)=0, \tag{1.2.12}$$

where l is an operator of the first order, A and B are second order operators, and A is an elliptic-parabolic operator.

Thus, equation (1.2.1) is either a hyperbolic equation or a composite type equation.

For third order equations with more than two independent variables, we do not have such a clear partition into the types dependent on the coefficients. Evidently, the number of types for third order equations with many independent variables (more than two) is essentially greater than for the third order equations with two independent variables. Among the third order equations, we select the equations that may be represented in the form (1.2.12). Then, all the equations (1.2.2), (1.2.5)-(1.2.11) will enter this class. In other words, the class of equations represented in the form (1.2.12) will contain the classical equations of the composite type and the parabolic-composite type, pseudoparabolic and pseudohyperbolic equations, and some other equations.

Thus, the equations represented in the form (1.2.12), we select as the important canonical class of equations of the composite type of the third order. This class of equations will be a matter of enquiry in our book.

In parallel with the common representation, these equations have the common methods for their investigation.

It should be noted that not all third order equations with more than two variables and which have both real and complex characteristics may be represented in form (1.2.12). Some of these equations may be solved by the methods given in this book and some of them, not.

1.3. BOUNDARY–VALUE PROBLEMS FOR THIRD ORDER EQUATIONS OF COMPOSITE TYPE

In a bounded domain D from the space R^n, we consider the equation

$$Lu \equiv lAu + Bu = f(x), \tag{1.3.1}$$

where A and B are the following operators of the second order

$$Au = a^{ij}(x)u_{x_i x_j} + a^i(x)u_{x_i} + a(x)u,$$

$$Bu = b^{ij}(x)u_{x_i x_j} + b^i(x)u_{x_i} + b(x)u,$$

and l is the operator of the first order

$$l = \alpha^k(x)\frac{\partial}{\partial x_k} + \alpha(x).$$

Here and further we assume that we sum by the repeating indices (the limits of summarizing will be indicated, if necessary). The aim of the monograph is to obtain the sufficient conditions for well-posedness of boundary-value problems; therefore, we shall not require the minimal smoothness conditions for the coefficients and the boundary. We shall assume that the functions

$$a^{ij}(x), a^i(x), \ldots$$

are infinitely differentiable in $\bar{D}$, and the boundary of D is a compact set and is piecewise smooth (i.e. belongs to the class $B^{k,\beta}$ with sufficiently large k). The definition of classes $B^{k,\beta}$ may be taken from Oleinik and Radkevich (1969).

We shall often use the notion of regular solutions. The regular solution of a certain boundary-value problem we shall define as a function which has in the given domain all the square integrable derivatives entering the equation, satisfying the equation, and attaining the boundary conditions on the average (we shall amend the notion of a regular solution if necessary).

Note that the minimal smoothness conditions may always be written starting from the proof of the concrete theorem.

We denote by Γ the boundary of the domain D: at the point of smoothness x, we define $\nu(x) = (\nu_1(x), \ldots, \nu_n(x))$ the external normal vector.

Further, in some cases, for brevity, we shall omit the dependence of coefficients of equations on the functions or on the variable x.

1.3.1. Hyperbolic boundary-value problem. Uniqueness of regular solutions

We suppose that an operator A is elliptic-parabolic, i.e. we suppose that

$$a^{ij}\xi_i\xi_j \geq 0, \qquad \forall x \in \bar{D}, \qquad \forall \xi \in R^n.$$

We shall make now the partition of smooth components of the boundary of D.

Taking account of the coefficients of the operator l, we shall select the sets σ_1, σ_2, and σ_0

$$\sigma_1 = \left\{x \in \Gamma : \alpha^k \nu_k > 0\right\}, \quad \sigma_2 = \left\{x \in \Gamma : \alpha^k \nu_k < 0\right\},$$

$$\sigma_0 = \left\{x \in \Gamma : \alpha^k \nu_k = 0\right\}.$$

From the sets σ_1, σ_2, and σ_0, we select the point sets which are characteristic and noncharacteristic relative to the operator A

$$\sigma_1^0 = \left\{x \in \sigma_1 : a^{ij} \nu_i \nu_j = 0\right\}, \quad \sigma_{10} = \left\{x \in \sigma_1 : a^{ij} \nu_i \nu_j > 0\right\},$$

$$\sigma_2^0 = \left\{x \in \sigma_2 : a^{ij} \nu_i \nu_j = 0\right\}, \quad \sigma_{20} = \left\{x \in \sigma_2 : a^{ij} \nu_i \nu_j > 0\right\},$$

$$\sigma_0^0 = \left\{x \in \sigma_0 : a^{ij} \nu_i \nu_j = 0\right\}, \quad \sigma_{00} = \left\{x \in \sigma_0 : a^{ij} \nu_i \nu_j > 0\right\}.$$

Further, from the sets σ_1^0, σ_2^0, and σ_0^0, we select the sets where the Fichera function of the operator A (Fichera, 1960; Oleinik and Radkevich, 1969) has a definite sign

$$\sigma_1^{01} = \left\{x \in \sigma_1^0 : \left(a^{ij}_{x_j} - a^i\right)\nu_i > 0\right\}, \quad \sigma_1^{02} = \left\{x \in \sigma_1^0 : \left(a^{ij}_{x_j} - a^i\right)\nu_i < 0\right\},$$

$$\sigma_1^{00} = \left\{x \in \sigma_1^0 : \left(a^{ij}_{x_j} - a^i\right)\nu_i = 0\right\},$$

$$\sigma_2^{01} = \left\{x \in \sigma_2^0 : \left(a^{ij}_{x_j} - a^i\right)\nu_i > 0\right\}, \quad \sigma_2^{02} = \left\{x \in \sigma_2^0 : \left(a^{ij}_{x_j} - a^i\right)\nu_i < 0\right\},$$

$$\sigma_2^{00} = \left\{x \in \sigma_2^0 : \left(a^{ij}_{x_j} - a^i\right)\nu_i = 0\right\},$$

$$\sigma_0^{01} = \left\{x \in \sigma_0^0 : \left(a^{ij}_{x_j} - a^i\right)\nu_i > 0\right\}, \quad \sigma_0^{02} = \left\{x \in \sigma_0^0 : \left(a^{ij}_{x_j} - a^i\right)\nu_i < 0\right\},$$

$$\sigma_0^{00} = \left\{ x \in \sigma_0^0 : \left(a^{ij}_{x_j} - a^i \right) \nu_i = 0 \right\}.$$

Now, we define the operator C as follows:

$$Cu = \left(\alpha^k a^{ij}_{x_k} - \alpha^i_{x_k} a^{kj} - a^{ik} \alpha^j_{x_k} + \alpha a^{ij} + b^{ij} \right) u_{x_i x_j}$$

$$+ \left(\alpha^k a^i_{x_k} - a^k \alpha^i_{x_k} - \alpha^i_{x_k x_j} a^{kj} + \alpha a^i + b^i \right) u_{x_i}$$

$$+ \left(\alpha^k a_{x_k} + \alpha a + b \right) u \equiv c^{ij} u_{x_i x_j} + c^i u_{x_i} + cu.$$

We suppose that the operator C is elliptic-parabolic

$$c^{ij} \xi_i \xi_j \geq 0, \qquad \forall x \in \bar{D}, \qquad \forall \xi \in R^n.$$

From the sets $\sigma_2^{02} \cup \sigma_2^{00}$ and $\sigma_0^{02} \cup \sigma_0^{00}$ we select the sets which are characteristic and noncharacteristic respectively relative to the operator C

$$\gamma_2^0 = \left\{ x \in \sigma_2^{02} \cup \sigma_2^{00} : c^{ij} \nu_i \nu_j = 0 \right\}, \quad \gamma_{20} = \left\{ x \in \sigma_2^{02} \cup \sigma_2^{00} : c^{ij} \nu_i \nu_j > 0 \right\},$$

$$\gamma_0^0 = \left\{ x \in \sigma_0^{02} \cup \sigma_0^{00} : c^{ij} \nu_i \nu_j = 0 \right\}, \quad \gamma_{00} = \left\{ x \in \sigma_0^{02} \cup \sigma_0^{00} : c^{ij} \nu_i \nu_j > 0 \right\}.$$

Boundary-value problem I: *Find solutions of equation (1.3.1) satisfying the following conditions*

$$u\Big|_{\sigma_1 \cup \sigma_{00} \cup \sigma_0^{01} \cup \gamma_{00}} = 0, \tag{1.3.2}$$

$$l_0 u\Big|_{\sigma_{10} \cup \sigma_1^{01} \cup \sigma_{20} \cup \sigma_2^{01} \cup \gamma_{20}} = 0, \tag{1.3.3}$$

where

$$l_0 u = \alpha^k u_{x_k}.$$

Before we study the boundary-value problem I, we note that this problem completely coincides with the first boundary-value problem for elliptic-parabolic equations (Fichera, 1960; Oleinik and Radkevich, 1969). Really, if we set

$$\alpha^k \equiv 0, \qquad \alpha \equiv 1, \qquad B \equiv 0,$$

then the smooth component of Γ will coincide with the set σ_0, and condition (1.3.2) will become the boundary condition (Fichera, 1960; Oleinik and Radkevich, 1969). Further, we may verify that for pseudoparabolic and pseudohyperbolic equations the boundary-value problem I is the first initial boundary-value problem (in the case of the cylindrical domain). Finally, for equations with the indeterminate characteristic form, the suggested boundary-value problem coincides with the problem studied by Vragov (1977). As an example, we consider the equation

$$K(x_0, x_1)u_{x_0x_0} - K^*(x_0, x_1)u_{x_1x_1} + a(x_0, x_1)u_{x_0}$$

$$+b(x_0, x_1)u_{x_1} + c(x_0, x_1)u = f(x_0, x_1),$$

where $K^*(x_0, x_1) \geq 0$ and the sign of $K(x_0, x_1)$ is not fixed.

We rewrite this equation in the form

$$\frac{\partial}{\partial x_0}\Big(- Ku_{x_0} - au\Big) + K^* u_{x_1x_1} + K_{x_0}u_{x_0} - bu_{x_1} - \Big(c - a_{x_0}\Big)u = f.$$

Evidently, this equation is of the form (1.3.1). Analyzing the boundary-value problem setting (Vragov, 1977) for equations with indeterminate characteristic form and comparing it with the setting of the boundary-value problem I, we find that they coincide.

Parallel with the boundary-value problem I, we may suggest a somewhat modified problem.

We determine the operator $\tilde{C}$ as follows:

$$\tilde{C}u = \Big(\alpha^k a^{ij}_{x_k} - \alpha^i_{x_k} a^{kj} - a^{ik}\alpha^j_{x_k} + b^{ij}\Big)u_{x_ix_j}$$

$$+\Big(\alpha^k a^i_{x_k} - a^k \alpha^i_{x_k} - \alpha^i_{x_kx_j} a^{ij} + b^i - \alpha_{x_k}(a^{ki} + a^{ik})\Big)u_{x_i}$$

$$+\left(\alpha^k a_{x_k} - \alpha_{x_i x_j} a^{ij} + b\right) u \equiv \tilde{c}^{ij} u_{x_i x_j} + \tilde{c}^i u_{x_i} + \tilde{c} u,$$

and suppose that this operator is an elliptic-parabolic operator.

By analogy with the sets γ_{20}, γ_2^0, γ_{00}, and γ_0^0, we introduce the sets $\tilde{\gamma}_{20}$, $\tilde{\gamma}_2^0$, $\tilde{\gamma}_{00}$, and $\tilde{\gamma}_0^0$.

Boundary-value problem I′: *Find the solutions of equation (1.3.1) satisfying the conditions*

$$u\Big|_{\sigma_1 \cup \sigma_{00} \cup \sigma_0^{01} \cup \tilde{\gamma}_{00}} = 0, \tag{1.3.2$'$}$$

$$lu\Big|_{\sigma_{10} \cup \sigma_1^{01} \cup \sigma_{20} \cup \sigma_2^{01} \cup \tilde{\gamma}_{20}} = 0. \tag{1.3.3$'$}$$

Evidently, the boundary-value problem I′ is also the generalization of the first boundary-value problem for elliptic-parabolic equations.

Existence and uniqueness of solutions of the boundary-value problems I and I′ are closely connected with solvability of certain boundary-value problems for corresponding first order equations. In particular, for uniqueness of regular solutions of the boundary-value problem I (I′), it is required that (side by side with certain conditions and the condition (I) (I′)) the domain D and the coefficients of the operator l be such that the problem

$$-\frac{1}{2}\left(\alpha^k \varphi\right)_{x_k} = 1, \qquad \varphi\Big|_{\sigma_2} = 0,$$

respectively,

$$-\frac{1}{2}\left(\alpha^k \varphi\right)_{x_k} + \alpha\varphi = 1, \qquad \varphi\Big|_{\sigma_2} = 0$$

has a regular solution in D.

Below we shall give the sufficient conditions providing the existence of the regular solution.

We introduce the following notations:

$$a_0(x) = -a(x) + \frac{1}{2} a^i_{x_i}(x) - \frac{1}{2} a^{ij}_{x_i x_j}(x),$$

$$d^{ij}(x) = \left(c^{ik}(x)\alpha^j(x)\right)_{x_k} - \frac{1}{2}\left(\alpha^k(x)c^{ij}(x)\right)_{x_k} - c^i(x)\alpha^j(x),$$

$$\tilde{d}^{ij}(x) = \left(\tilde{c}^{ik}(x)\alpha^j(x)\right)_{x_k} - \frac{1}{2}\left(\alpha^k(x)\tilde{c}^{ij}(x)\right)_{x_k} - c^i(x)\alpha^j(x) + \alpha(x)\tilde{c}^{ij}(x),$$

$$c_0(x) = \frac{1}{2}\left(\alpha^i(x)\tilde{c}^{ij}(x)\right)_{x_i} - \frac{1}{2}\left(\alpha(x)\tilde{c}^{ij}(x)\right)_{x_i x_j} + \frac{1}{2}\left(\alpha(x)\tilde{c}^i(x)\right)_{x_i} - \alpha(x)\tilde{c}(x).$$

Further, we shall denote by $\bar{a}_0$, $\bar{\alpha}_0$, α_0, δ, and δ_0 certain constants.

To establish uniqueness of regular solutions of the boundary-value problem I, we shall require one of the following conditions

(2) $a^{ij}\xi_i\xi_j \geq \delta|\xi|^2$, $a_0(\alpha^i\xi_i)^2 + d^{ij}\xi_i\xi_j \geq 0$, $\delta > 0$, $x \in \bar{D}$, $\xi \in R^n$

and that one of the sets $\sigma_{00} \cup \sigma_0^{01} \cup \gamma_{00}$ or $\sigma_{10} \cup \sigma_1^{01} \cup \sigma_2^{01} \cup \gamma_{20}$ is not empty;

(3) $a^{ij}\xi_i\xi_j \geq 0$, $a_0(\alpha^i\xi_i)^2 + d^{ij}\xi_i\xi_j \geq \delta|\xi|^2$, $\delta > 0$, $x \in \bar{D}$, $\xi \in R^n$

and that the set $\sigma_1 \cup \sigma_{00} \cup \sigma_0^{01} \cup \gamma_{00}$ is not empty;

(4) $a^{ij}\xi_i\xi_j \geq 0$, $d^{ij}\xi_i\xi_j \geq 0$, $a_0 \geq \bar{a}_0 > 0$, $x \in \bar{D}$, $\xi \in R^n$;

(5) $a^{ij}\xi_i\xi_j \geq 0$, $a_0(\alpha^i\xi_i)^2 + d^{ij}\xi_i\xi_j \geq 0$, $(\alpha^i c)_{x_i} \geq \delta > 0$, $x \in \bar{D}$, $\xi \in R^n$.

Theorem 1.3.1. *Let the operator A and C be elliptic-parabolic and symmetric and suppose that the following conditions*

$$\sum_{i=1}^{n}\left[\alpha^i(x)\right]^2 \not\equiv 0, \quad \left(\alpha^i c\right)_{x_i} \geq 0, \quad \text{for } x \in \bar{D}, \quad c \geq 0, \quad \text{for } x \in \sigma_2,$$

hold. Let, besides, either the conditions **(1)** *and* **(2)**, *or the condition* **(3)** *or the conditions* **(1)** *and* **(4)**, *or the condition* **(5)** *hold.*

Then the regular solution of the boundary-value problem I is unique.

Proof. Consider the equality

$$-\int\limits_D Lu \cdot l_0 u dx = -\int\limits_D f \cdot l_0 u dx. \tag{1.3.4}$$

Integrating by parts, we may transform the left side of this equality to the shape

$$\int\limits_D \left[a^{ij}(l_0u)_{x_i}(l_0u)_{x_j} + a_0(l_0u)^2 + d^{ij}u_{x_i}u_{x_j} + \frac{1}{2}(\alpha^i c)_{x_i}u^2\right] dx$$

$$-\frac{1}{2}\int\limits_\Gamma \left(a^{ij}_{x_j} - a^i\right)\nu_i(l_0u)^2 d\gamma + \int\limits_\Gamma a^{ij}(l_0u)_{x_i} l_0 u \nu_j d\gamma$$

$$-\frac{1}{2}\int\limits_\Gamma \alpha^k \nu_k c^{ij} u_{x_i}u_{x_j} d\gamma + \int\limits_\Gamma \alpha^k u_{x_k} c^{ij} u_{x_i}\nu_j d\gamma + \frac{1}{2}\int\limits_\Gamma \alpha^k \nu_k c u^2 d\gamma.$$

By the hypotheses of the theorem, the integral over the domain D in this expression is estimated from below by the value

$$k_1\int\limits_D \left(\sum_{i=1}^n \left[\left(l_0 u\right)_{x_i}\right]^2\right) dx + k_2\int\limits_D (l_0u)^2 dx + k_3\int\limits_D \left(\sum_{i=1}^n u^2_{x_i}\right) dx + k_4\int\limits_D u^2 dx,$$

where one of the constants k_1, k_2, k_3, k_4 is positive, and the rest are equal to zero.

We consider now the boundary integrals which arise in the transformed left side of equality (1.3.4).

First, we note that at points of the set $\sigma_{01} \cup \sigma_1^{01}$ we have $u_{x_k} = 0$, $k = 1, \dots, n$. Really, the unit normal vector ν has at most one of the non-zero coordinate ν_i. Suppose, for definiteness, at the point $x \in \sigma_{10} \cup \sigma_1^{01}$, we have $\nu_1 \neq 0$. The vectors

$$\left(\nu_2, -\nu_1, 0, \dots, 0\right), \left(\nu_3, 0, -\nu_1, 0, \dots, 0\right), \dots \left(\nu_n, 0, \dots, 0, -\nu_1\right)$$

are lying in the plane tangent to Γ in this point. Since the tangent derivatives are equal to zero and $l_0 u = 0$ at the point x, we obtain the system for the derivatives $u_{x_1}, \ldots, u_{x_n}$

$$\alpha^1 u_{x_1} + \ldots + \alpha^n u_{x_n} = 0,$$

$$\nu_2 u_{x_1} - \nu_1 u_{x_2} = 0,$$

$$\cdots\cdots$$

$$\nu_n u_{x_1} - \nu_1 u_{x_n} = 0, \tag{1.3.5}$$

whose determinant is equal to

$$(-1)^{n-1} \nu_1^{n-2} (\alpha^k \nu_k).$$

As this determinant is not equal to zero, system (1.3.5) may have only a trivial solution. This means that $u_{x_k} = 0$, $k = 1, \ldots, n$.

Further, we note that at the points of the set $\sigma_{00} \cup \sigma_0^{01} \cup \gamma_{00}$ we have $l_0 u = 0$. Really, suppose that x is an arbitrary point of the set $\sigma_{00} \cup \sigma_0^{01} \cup \gamma_{00}$. In this point, either all the partial derivatives $u_{x_1}, \ldots, u_{x_n}$ are equal to zero, or there exists at most one nontrivial derivative. If all the partial derivatives are equal to zero, then we shall have $l_0 u = 0$. Suppose that at most one of the partial derivatives of the function $u(x)$ at the point x is not equal to zero. Suppose, for definiteness, it will be u_{x_1}. Taking into account that the tangent derivatives at the point x are equal to zero and the definition of σ_0, we obtain the following system

$$\alpha^1 \nu_1 + \ldots + \alpha^n \nu_n = 0,$$

$$u_{x_2} \nu_1 - u_{x_1} \nu_2 = 0,$$

$$u_{x_n}\nu_1 - u_{x_1}\nu_n = 0.$$

We have obtained the linear homogeneous system relative to the coordinates of the vector ν. The determinant of this system is equal to

$$(-1)^{n-1}u_{x_1}^{n-2}(\alpha^k u_{x_k}).$$

As $|\nu| = 1$, then this determinant must be equal to zero. Therefore, in the point x, we shall have $\alpha^k u_{x_k} = 0$, which was to be proved.

Finally, we may note that at the points of the sets σ_1^0, σ_2^0, and σ_0^0 the equalities

$$a^{ij}\nu_j = 0, \qquad i = 1, \dots, n$$

hold; and at the points of the sets γ_2^0 and γ_0^0 the equalities

$$c^{ij}\nu_j = 0, \qquad i = 1, \dots, n$$

hold. The proof of these statements is known and may be found in (Fichera, 1960; Oleinik and Radkevich, 1969).

All that we have established and the equalities

$$u_{x_j}\nu_k = u_{x_k}\nu_j, \qquad j, k = 1, \dots, n,$$

which hold for the points of the sets σ_1^{02} and σ_1^{00} (these equalities again give vanishing tangent derivatives), allows us to state that the boundary integrals, which arise in the transformed left side of equality (1.3.4), are nonnegative.

We suppose now that $f \equiv 0$. If the condition (**5**) holds, then the number k_4 will be positive and the calculations above will give the inequality

$$\int\limits_D u^2 dx \leq 0,$$

whence we obtain that $u \equiv 0$ in D. If the condition (**3**) holds, then the number k_3 will be positive, and the inequality

$$\int_D \left(\sum_{i=1}^{n} u_{x_i}^2 \right) dx \le 0$$

will hold true, whence we obtain that $u(x) = \text{const}$ in D. As $u(x)$ vanishes in the nonempty subset of the boundary, we obtain again that $u \equiv 0$ in D.

Now, we suppose that the conditions (**1**) and (**4**) hold. Then, the number k_2 will be positive and the inequality

$$\int_D (l_0 u)^2 dx \le 0$$

will hold, whence we obtain $l_0 u \equiv 0$ in D. Now, we multiply this identity from the right and from the left by the function φu where φ is the function defined in the condition (**1**), and integrate it over the domain D. As a result, we obtain the equality

$$\int_D u^2 dx \equiv 0,$$

which gives again $u \equiv 0$ in D.

Finally, we suppose that the conditions (**1**) and (**2**) hold. Then, the following inequality

$$\int_D \left(\sum_{i=1}^{n} \left[(l_0 u)_{x_i} \right]^2 \right) dx \le 0$$

will hold true, whence we obtain that $lu \equiv \text{const}$ in D. If the set

$$\sigma_{10} \cup \sigma_1^{01} \cup \sigma_{20} \cup \sigma_2^{01} \cup \gamma_{20}$$

is not empty, we immediately obtain that $l_0 u \equiv 0$ in D. Again, using the condition (**1**), we may show that $u \equiv 0$ in D. If the set $\sigma_{00} \cup \sigma_0^0 \cup \gamma_{00}$ is not empty, then, as was mentioned above, we shall have $l_0 u = 0$ in this set. Repeating the previous considerations, we obtain again that $u \equiv 0$ in D. The theorem is proved.

To establish uniqueness of the regular solutions of the boundary-value problem I′, we shall need the following conditions

(2′) $a^{ij}\xi_i\xi_j \geq \delta|\xi|^2$, $a_0(\alpha^i\xi_i + \alpha\xi_0)^2 + \tilde{d}^{ij}\xi_i\xi_j + c_0\xi_0^2 \geq 0$,
$\delta > 0$, $x \in \bar{D}$, $\xi = (\xi_1, \dots, \xi_n) \in R^n$, $\xi_0 \in R$, $i, j = 1, \dots, n$

one of the sets $\sigma_{00} \cup \sigma_0^{01} \cup \tilde{\gamma}_{00}$ or $\sigma_{10} \cup \sigma_1^{01} \cup \sigma_2^{01} \cup \tilde{\gamma}_{20}$ is not empty;

(3′) $a^{ij}\xi_i\xi_j \geq 0$, $a_0(\alpha^i\xi_i + \alpha\xi_0)^2 + \tilde{d}^{ij}\xi_i\xi_j + c_0\xi_0^2 \geq \delta|\xi|^2$,
$\delta > 0$, $x \in \bar{D}$, $\xi \in R^n$, $\xi_0 \in R$

the set $\sigma_1 \cup \sigma_{00} \cup \sigma_0^{01} \cup \tilde{\gamma}_{00}$ is not empty;

(4′) $a^{ij}\xi_i\xi_j \geq 0$, $d^{ij}\xi_i\xi_j \geq 0$, $c_0 \geq 0$, $a_0 \geq \bar{a}_0 > 0$,
$x \in \bar{D}$, $\xi \in R^n$, $\xi_0 \in R$;

(5′) $a^{ij}\xi_i\xi_j \geq 0$, $a_0(\alpha^i\xi_i + \alpha\xi_0)^2 + \tilde{d}^{ij}\xi_i\xi_j + c_0\xi_0^2 \geq \delta\xi_0^2$,
$\delta > 0$, $x \in \bar{D}$, $\xi \in R^n$, $\xi_0 \in R$.

Theorem 1.3.1′. *Let the operators A and C be elliptic-parabolic and symmetric, and suppose that the condition*

$$\left[\alpha c^i - (\alpha\tilde{c}^{ij})_{x_j} + \alpha^i c\right]\nu_i \geq 0 \quad \text{for} \quad x \in \sigma_2 \cup \tilde{\gamma}_0^0$$

holds and either the conditions **(1′)**, **(2′)**, *or the condition* **(3′)**, *or the conditions* **(1′)**, **(4′)**, *or the condition* **(5′)** *hold.*

Then the regular solution of the boundary-value problem I' is unique.

The proof of Theorem 1.3.1′ is realized by the same scheme as for Theorem 1.3.1; only we analyze the equality

$$-\int_D Lu \cdot lu dx = -\int_D f \cdot lu dx.$$

Remark. Generally speaking, to prove Theorems 1.3.1 and 1.3.1′ it suffices to require that the operators C and $\tilde{C}$ are elliptic-parabolic only in the set $\sigma_1 \cup \sigma_2 \cup \sigma_0^{02} \cup \sigma_0^{00}$.

1.3.2. Existence of the regularized solutions of the hyperbolic boundary-value problem

Before establishing the existence of solutions of the boundaryvalue problems I and I′, we shall formulate some results on the properties of the boundary-value problem solutions for the first order equations. Further we shall need these results and, besides, they have their own importance.

Consider now the boundary-value problem

$$l_0 u = w, \qquad u\Big|_{\sigma_1} = 0. \tag{1.3.6}$$

Suppose that the set σ_0 consists of m connected components $\sigma_{0,1}, \dots, \sigma_{0,m}$ of dimension $n-1$ and the set $\sigma_{0,m+1}$ of dimension less than $n-1$. We suppose, besides, each component $\sigma_{0,i}$, $i = 1, \dots, m$ is described by the equation $F_i(x) = 0$, where $F_i(x)$ is a smooth function. We make the change of variables $y = \psi_i(x)$ for ith component, which transforms this component into the subset of the plane $y_n = 0$ (for this transformation we have $y_n = F_i(x)$). We suppose that the transformation ψ_i is one-to-one in $\sigma_{0,i}$, and that it may be extended to a one-to-one continuous mapping in the set $\bar{\sigma}_{0,i}$. The set $\sigma_{0,i}$ under this mapping, will be transformed into a certain set from the plane $y_n = 0$; the operator l_0 will be transformed into the operator

$$\tilde{l}_0 = \sum_{k=1}^{n-1} \tilde{\alpha}^k \frac{\partial}{\partial y_k} + \left(\alpha^k \frac{\partial F_i}{\partial x_k}\right) \frac{\partial}{\partial y_n}.$$

As, in the set $\sigma_{0,i}$, we have

$$\alpha^k \frac{\partial F_i}{\partial x_k} = 0,$$

then, in this set, the operator l_0 is transformed into the operator $\tilde{l}_0$ with respect to the variables $y_1, \dots, y_{n-1}$ with the coefficients dependent only on variables $y_1, \dots, y_{n-1}$. Analogous to the sets σ_1, σ_2, σ_0, we define the sets

$$\sigma_{0,i,1}(\tilde{l}_0), \quad \sigma_{0,i,2}(\tilde{l}_0), \quad \sigma_{0,i,0}(\tilde{l}_0)$$

in the boundary of the image of the set $\sigma_{0,i}$.

We suppose that the subset $\tilde{\sigma}_0$ of the set σ_0 consists of the sets $\sigma_{0,1}$, $\sigma_{0,2}$, …, σ_{0,m_0} of the dimension $n-1$ and a certain set of the dimension less then

$n-1$. Moreover, we suppose that the following conditions hold for the sets $\sigma_{0,1}, \sigma_{0,2}, \dots, \sigma_{0,m_0}$:

$$\sigma_{0,1,1}(\tilde{l}_0) \subset \psi_1\Big(\partial\sigma_1 \bigcap \bar{\sigma}_{0,1}\Big),$$

$$\sigma_{0,2,1}(\tilde{l}_0) \subset \psi_1\Big(\partial\sigma_{0,1}\Big) \bigcup \psi_2\Big(\partial\sigma_1 \bigcap \bar{\sigma}_{0,2}\Big),$$

$$\sigma_{0,3,1}(\tilde{l}_0) \subset \psi_1\Big(\partial\sigma_{0,1}\Big) \bigcup \psi_2\Big(\partial\sigma_{0,2}\Big) \bigcup \psi_3\Big(\partial\sigma_1 \bigcap \bar{\sigma}_{0,3}\Big), \tag{1.3.7}$$

$$\dots\dots\dots$$

and similar to the set $\sigma_{0,m_0,1}(\tilde{l}_0)$.

Lemma 1.3.1. *For a certain natural k, $1 \le k \le n$, we have*

$$\Big|\alpha^k(x)\Big| \ge \delta > 0 \quad \text{for} \quad x \in \bar{D},$$

and suppose that the intersection of the sets $\bar{\sigma}_1$ and $\overline{\sigma_0 \cup \sigma_2}$ satisfies the cone condition. If a function $w(x)$ belongs to the space $W_2^1(D)$ and vanishes in the subset $\tilde{\sigma}_1$ of the set σ_1, and if in the set $\sigma_1 \backslash \tilde{\sigma}_1$ the condition

$$\alpha^k \nu_k \ge \bar{\alpha}_0 > 0$$

holds, then problem (1.3.6) will have a solution from the space $W_2^1(D)$, which will satisfying the following estimate

$$\sum_{i=1}^{n}\left(\int\limits_{\sigma_1\backslash\tilde{\sigma}_1} \alpha^k \nu_k v_{x_i}^2\, d\gamma - \int\limits_{\sigma_2} \alpha^k \nu_k v_{x_i}^2\, d\gamma\right)$$

$$-\int\limits_{\sigma_2} \alpha^k \nu_k v^2\, d\gamma + \Big|\Big|v\Big|\Big|^2_{W_2^1(D)} \le R_0 \Big|\Big|w\Big|\Big|^2_{W_2^1(D)} \tag{1.3.8}$$

with the constant R_0 dependent only on the coefficients of the operator

l_0 and the domain D. If the function $w(x)$ belongs to the space $W_2^2(D)$, vanishes in the set σ_1 and the condition

$$\alpha^k \nu_k \geq \bar{\alpha}_0 > 0 \quad \text{for} \quad x \in \sigma_1$$

holds, then problem (1.3.6) will have a solution from the space $W_2^2(D)$ satisfying the estimate

$$\left|\left|u\right|\right|_{W_2^2(D)} \leq R_0 \left|\left|w\right|\right|_{W_2^2(D)}, \tag{1.3.9}$$

with the constant R_1 dependent only on the coefficients of the operator l_0 and the domain D. Finally, if for the subset $\tilde{\sigma}_0$ of the set σ_0, condition (1.3.7) holds, and if the function $w(x)$ vanishes for $x \in \tilde{\sigma}_0$ then the function $u(x)$ will vanish for $x \in \tilde{\sigma}_0$.

Proof. For $\varepsilon > 0$, we consider the problem: *find the solution of the equation*

$$\Phi_\varepsilon u \equiv -\varepsilon \Delta u + l_0 u = w, \tag{1.3.10}$$

satisfying the conditions

$$u\Big|_{\sigma_1} = 0, \qquad \frac{\partial u}{\partial \nu}\Big|_{\sigma_0 \cup \sigma_2} = 0. \tag{1.3.11}$$

By the hypotheses of the lemma, this problem has the solution $u^\varepsilon(x)$ from the space $W_2^2(D)$ (see Kerimov, 1977). We suppose that, for definiteness, the following condition

$$\alpha^1(x) \geq \delta > 0 \quad \text{for} \quad x \in \bar{D}$$

holds. Rather simple analysis of the equality

$$\int_D \Phi_\varepsilon u^\varepsilon \exp(-\mu x_1) dx = \int_D w u^\varepsilon \exp(-\mu x_1) dx,$$

where μ is a positive number, gives the first *a priori* estimate

$$\varepsilon\sum_{i=1}^{n}\int\limits_D \left(u^\varepsilon_{x_i}\right)^2 dx + \int\limits_D \left(u^\varepsilon\right)^2 dx - \int\limits_{\sigma_2} \alpha^k\nu_k\left(u^\varepsilon\right)^2 d\gamma \leq K_1 \int\limits_D w^2 dx \quad (1.3.12)$$

with a constant K_1 independent of ε.

Now, we consider the equality

$$-\int\limits_D \Phi_\varepsilon u^\varepsilon \frac{\partial}{\partial x_i}\Big(\exp(-\mu x_1)u^\varepsilon_{x_i}\Big)dx = -\int\limits_D w\frac{\partial}{\partial x_i}\Big(\exp(-\mu x_1)u^\varepsilon_{x_i}\Big)dx.$$

Integrating by parts and taking into account boundary conditions (1.3.11), we may transform this equality as follows:

$$\int\limits_D \left\{\left[\frac{\mu}{2}\alpha^1 - \frac{1}{2}\alpha^k_{x_k}\right]\left(u^\varepsilon_{x_i}\right)^2 + \alpha^k_{x_i}u^\varepsilon_{x_k}u^\varepsilon_{x_i}\right\}\exp(-\mu x_1)dx$$

$$+\varepsilon\int\limits_D \left(\Delta u^\varepsilon\right)^2 \exp(-\mu x_1)dx - \mu\varepsilon\int\limits_D \Delta u^\varepsilon u^\varepsilon_{x_1}dx$$

$$+\frac{1}{2}\int\limits_{\sigma_1} \alpha^k\nu_k\left(u^\varepsilon_{x_i}\right)^2 \exp(-\mu x_1)d\gamma - \frac{1}{2}\int\limits_{\sigma_2} \alpha^k\nu_k\left(u^\varepsilon_{x_i}\right)^2 \exp(-\mu x_1)d\gamma$$

$$= \int\limits_D u^\varepsilon_{x_i}w_{x_i}\exp(-\mu x_1)dx + \int\limits_{\sigma_1\setminus\hat{\sigma}_1} u^\varepsilon_{x_i}w\exp(-\mu x_1)\nu_i d\gamma. \quad (1.3.13)$$

The boundary integral standing in the right side of this equality may be estimated taking into account the condition $\alpha^k\nu^k \geq \bar{\alpha}_0 > 0$ in the set $\sigma_1\setminus\hat{\sigma}_1$ and the Young inequality by the value

$$\rho\int\limits_{\sigma_1} \alpha^k\nu_k\left(u^\varepsilon_{x_i}\right)^2 d\gamma + c(\rho)\int\limits_\Gamma w^2 d\gamma.$$

Here ρ is an arbitrary positive number (here, as in equality (1.3.13) and in the formulas we shall obtain further, we sum by the indices k and i).

Now we may choose the positive constant μ so large that the quadratic form

$$\left[\frac{\mu}{2}\alpha^1 - \frac{1}{2}\alpha^k_{x_k}\right]\eta_i^2 + \alpha^k_{x_i}\eta_k\eta_i, \quad \eta \in R^n$$

will be positive definite. Applying to the first integral from the right side of equality (1.3.13) and to the first integral of the left side the Young inequality and choosing the constant ρ sufficiently small, we may easily obtain the inequality

$$\mu_0\left[\int_D \left(u^\varepsilon_{x_i}\right)^2 dx + \varepsilon\int_D \left(\Delta u^\varepsilon\right)^2 dx + \int_{\sigma_1} \alpha^k\nu_k\left(u^\varepsilon_{x_i}\right)^2 d\gamma\right.$$

$$\left.-\int_{\sigma_2} \alpha^k\nu_k\left(u^\varepsilon_{x_i}\right)^2 d\gamma\right] \le \frac{\varepsilon\rho_1^2}{2}\int_D \left(\Delta u^\varepsilon\right)^2 dx + \frac{\mu^2\varepsilon}{2\rho_1^2}\int_D \left(u^\varepsilon_{x_i}\right)^2 dx$$

$$+\frac{\rho_2^2}{2}\int_D \left(u^\varepsilon_{x_i}\right)^2 dx + \frac{1}{2\rho_2^2}\int_D w^2_{x_i} dx + K_2\int_\Gamma w^2 d\gamma.$$

In this inequality, the positive constants μ_0 and K_2 depend only on coefficients of the operator l_0 and on the domain D; the positive constants ρ_1 and ρ_2 are arbitrary. Choosing these two constants arbitrarily small, using estimate (1.3.12) and the inequality

$$\left|\left|w\right|\right|_{L_2(\Gamma)} \le K_0\left|\left|w\right|\right|_{W_2^1(D)}$$

(see Sobolev, 1950; Ladyzhenskaya and Ural'tseva, 1973) we may easily obtain the estimate

$$\varepsilon\int_D \left(\Delta u^\varepsilon\right)^2 dx + \left|\left|u^\varepsilon\right|\right|^2_{W_2^1(D)} + \int_{\sigma_1} \alpha^k\nu_k\left(u^\varepsilon_{x_i}\right)^2 d\gamma$$

$$-\int_{\sigma_2} \alpha^k\nu_k\left(u^\varepsilon_{x_i}\right)^2 d\gamma \le K_2\left|\left|w\right|\right|^2_{W_2^1(D)}. \qquad (1.3.14)$$

Passing now in equation (1.3.10) to the limit for $\varepsilon \to 0$, we may prove in a standard way that there exists such a limit function $u(x)$ which solves problem (1.3.6). This solution will satisfy inequalities (1.3.12) and (1.3.14) taken for $\varepsilon = 0$. These inequalities and the equality $u_{x_i} = 0$ for $x \in \tilde{\sigma}_1$ denote that for the solution of problem (1.3.6) the estimate (1.3.8) will hold.

We shall prove now the second part of the lemma. First, we suppose that the solution $u(x)$ of problem (1.3.6) is a smooth function. We differentiate the equation $l_0 u = w$ with respect to the variable x_j. Note that the derivatives $u_{x_1}, \dots, u_{x_n}$ will now vanish in the whole set σ_1. In this case, for the second derivatives of the function $u(x)$ at the points of the set σ_1, we may write down the systems analogous to system (1.3.5). Only, in the right side of the first equation, the derivative w_{x_j} will stand. Calculating the values of these second derivatives and taking into account the condition $\alpha^k \nu_k \geq \bar{\alpha}_0 > 0$ in the set σ_1, we obtain the estimate

$$\int\limits_{\sigma_1} u^2_{x_i x_j} d\gamma \leq K_3 \int\limits_{\Gamma} w^2_{x_i} d\gamma. \tag{1.3.15}$$

Here the constant K_3 depends only on the coefficients of the operator l_0; the index j is fixed; and we sum by the index i.

Now, we differentiate the equation $l_0 u = w$ with respect to the variables x_i and x_j, multiply the equation obtained from the left and right by the expression $u_{x_i x_j} \exp(-\mu x_1)$ and integrate over the domain D. Using estimate (1.3.15) and the inequality (Sobolev, 1950; Ladyzhenskaya and Ural'tseva, 1973)

$$\int\limits_{\Gamma} w^2_{x_i} d\gamma \leq \tilde{K}_3 \left|\left| w \right|\right|^2_{W_2^2(D)},$$

we easily obtain the second estimate of the lemma, only for the smooth solutions of problem (1.3.6). However, the standard closure procedure allows us to state that the estimate obtained will denote that problem (1.3.6) has a solution from the space $W_2^2(D)$, and for this solution the established estimate holds.

Finally, we shall prove the third part of the lemma.

In the set $\psi(\sigma_{0,1})$, we have the equality $\tilde{l}_0 u = 0$. Moreover, the equality $u = 0$ holds in the set $\sigma_{0,1,1}(\tilde{l}_0)$. Therefore, from the first *a priori* estimate for the operator $\tilde{l}_0$ in the set $\psi(\sigma_{0,1})$, it follows that $u = 0$ in this set. Hence it follows that in the set $\sigma_{0,1}$ we shall have the equality $u = 0$. Then, in the set $\psi_2(\sigma_{0,2})$, we shall have the equality $\tilde{l}_0 u = 0$. From this and the second inclusion (1.3.7), it follows that the equality $u = 0$ will also hold in the set $\sigma_{0,2,1}(\tilde{l}_0)$. Again, taking account of the first *a priori* estimate for the operator $\tilde{l}_0$ in the set $\psi_2(\sigma_{0,2})$, we obtain that in this set the equality $u = 0$ will also hold. Hence, we again obtain the equality $u = 0$ for the points of the set $\sigma_{0,2}$.

In the second step we pass to the set $\sigma_{0,3}$ and so on. Making a finite number of steps, we exhaust all the set $\tilde{\sigma}_0$. The lemma is proved.

Remark 1. For the boundary-value problem

$$lu = w, \qquad u\Big|_{\sigma_1} = 0$$

we may establish the analogous lemma if we require that the condition

$$\alpha(x) \geq \alpha_0 > 0$$

holds for $x \in \bar{D}$, and for sufficiently large constant α_0.

Remark 2. Evidently, the sets σ_1 and σ_2 may be interchanged (with the corresponding change of conditions). The properties of solutions of equations $l_0 u = w$ or $lu = w$ with setting of the solution in the set σ_2 will be sufficient to satisfy the conditions (**1**) or (**1**$'$) of Theorems 1.3.1 and 1.3.1$'$.

Further, we shall suppose that the operator C is elliptic-parabolic in the set $\sigma_1 \cup \sigma_2 \cup \sigma_0^{02} \cup \sigma_0^{00}$. We define the sets γ_1 and γ_2 as follows:

$$\gamma_1 = \Big\{x \in \sigma_{20} \cup \sigma_2^{01} \cup \gamma_{20} : c^{ij}\nu_i\nu_j = 0\Big\},$$

$$\gamma_2 = \Big\{x \in \sigma_{20} \cup \sigma_2^{01} \cup \gamma_{20} : c^{ij}\nu_i\nu_j > 0\Big\}.$$

For brevity, we shall use the following notations

$$\Gamma_1 = \sigma_1^{01} \cup \sigma_2^{01} \cup \sigma_0^{01}, \qquad \Gamma_2 = \sigma_1^{02} \cup \sigma_2^{02} \cup \sigma_0^{02},$$

$$\Gamma_3 = \sigma_{10} \cup \sigma_{20} \cup \sigma_{00} \cup \sigma_0^{01} \cup \gamma_{20} \cup \gamma_{00}, \quad F(x) = \Big[a^{ij}_{x_j}(x) - a^i(x)\Big]\nu_i.$$

We suppose that W is the set of functions $w(x)$ from the space $W_2^2(D)$ equal to zero in the set Γ_3; W_1 is the set of functions $v(x)$ such that $l_0 v$ is an element of W and $v(x)$ vanishes in the set σ_1. Further, we shall use the space H, which we define as the closure of the set W_1 with respect to the norm

$$\left\|v\right\|_H = \Bigg\{ \int\limits_D \left[a^{ij}(l_0 v)_{x_i}(l_0 v)_{x_j} + (l_0 v)^2\right] dx + \int\limits_{\Gamma_1 \cup \Gamma_2} |F|(l_0 v)^2 d\gamma$$

$$+ \int\limits_{\sigma_1^0 \cup \sigma_2} \left|\alpha^k \nu_k\right| c^{ij} v_{x_i} v_{x_j} d\gamma \Bigg\}^{1/2}.$$

For the functions u from H and w from W, we define the bilinear form

$$a(u, w) = \int\limits_D \Bigg\{ -a^{ij}(l_0 u)_{x_i} w_{x_j} + \left(a^{ij}_{x_j} - a^i\right) l_0 u w_{x_i}$$

$$+\left(a - a^j_{x_i} + a^{ij}_{x_i x_j}\right) l_0 u w + u C^* w \Bigg\} dx + \int\limits_{\Gamma_2} F l_0 u w d\gamma$$

$$+ \int\limits_{\gamma_2} c^{ij} u w_{x_j} \nu_i d\gamma + \int\limits_{\sigma_2^{01} \cup \gamma_2^0} \left(c^{ij}_{x_j} - c^i\right) u w \nu_i d\gamma - \int\limits_{\sigma_1^{00} \cup \sigma_1^{01} \cup \sigma_1^{02}} c^{ij} u_{x_i} w \nu_j d\gamma,$$

where C^* is the operator conjugate to C.

Lemma 1.3.2. *Let, for a certain natural k, $1 \le k \le n$, the condition*

$$\left|\alpha^k(x)\right| \ge \delta_0 > 0 \quad \text{for} \quad x \in \bar{D}$$

hold and suppose that the intersection of sets $\bar{\sigma}_1$ and $\overline{\sigma_0 \cup \sigma_2}$ satisfies the cone condition. Let, besides, the condition

$$\left|\alpha^k \nu_k\right| \ge \bar{\alpha}_0 > 0 \quad \text{for} \quad x \in \sigma_1^0 \cup \gamma_2 \cup \sigma_2^{01} \cup \gamma_2^0$$

hold. Then for each of the functions $w(x)$ from the set W and $u(x)$ from the set W_1, we have the estimate

$$\left|a(u,w)\right| \leq M\left|\left|u\right|\right|_H,$$

where the constant M depends only on the coefficients of the operator L and the function $w(x)$.

Proof. As the hypotheses of Lemma 1.3.1 hold, for the functions $u(x)$ from the set W_1, we have the following estimate

$$-\int\limits_{s_2} \alpha^k \nu_k u^2 d\gamma + \int\limits_D u^2 dx \leq K_1 \int\limits_D (l_0 u)^2 dx \tag{1.3.16}$$

(see estimate (1.3.12)). From this estimate and the Cauchy – Schwarz – Bunyakovskii inequality, the required estimate follows for all the integrals taken over the domain D entering the form $a(u,w)$. Further, the integral estimate by the set Γ_2 follows from the Hölder inequality. The estimate of the integral by the set $\sigma_1^{00} \cup \sigma_1^{01} \cup \sigma_1^{02}$ follows from the Cauchy – Schwarz – Bunyakovskii inequality and from the condition

$$\alpha^k \nu_k \geq \bar{\alpha}_0 > 0$$

in this set. Finally, the estimate of the integrals by the sets γ_2 and $\sigma_2^{01} \cup \gamma_2^0$ is obtained taking account of the Hölder inequality, the condition

$$-\alpha^k \nu_k \geq \bar{\alpha}_0 > 0$$

and again the estimate (1.3.16).

These estimates give the required statement. The lemma is proved.

From Lemma 1.3.2 it follows that the form $a(u,w)$, for fixed w from the set W, may be extended from the set W_1 to the whole space H. We shall preserve for the extended form the notation $a(u,w)$.

Definition 1.3.1. *A function $u(x)$ from the space H is defined as the generalized solution of the boundary-value problem I, if for each function $w(x)$ from the set W the following identity holds*

$$a(u,w)=\int\limits_D fw\,dx.$$

Theorem 1.3.2. *Let operators A and C be symmetric, the operator A be elliptic-parabolic in $\bar{D}$, and the operator C be elliptic-parabolic in $\sigma_1\cup\sigma_2\cup\sigma_0^{02}\cup\sigma_0^{00}$. Suppose, besides, that the conditions of Lemma 1.3.2 hold; for the set $\sigma_{00}\cup\sigma_0^{01}\cup\gamma_{00}$, condition (1.3.7) holds; and, for the operator L coefficients, the following conditions hold*

$$d^{ij}\xi_i\xi_j\geq 0,\quad a_0\geq\bar{a}_0>0,\quad (\alpha^k c)_{x_k}\geq 0,\quad \text{for } x\in\bar{D},\ \xi\in R^n;$$

$$c^{ij}\nu_i\nu_j\equiv 0\quad\text{for}\quad x\in\sigma_2^{01};$$

$$\left(c_{x_j}^{ij}-c^i\right)\nu_i\equiv 0\quad\text{for}\quad x\in\gamma_0^0;$$

$$c\geq 0\quad\text{for}\quad x\in\sigma_2.$$

Then, for each function f from the space $L_2(D)$, there exists a regularized solution of the boundary-value problem I.

Proof. For a function u from the space H and a function v from the set W_1, we define the form $a_0(u,v)$ as follows

$$a_0(u,v)=a(u,l_0v).$$

As for a fixed function v, the form $a_0(u,v)$ is a linear continuous functional in the space H, then, by the Riesz theorem, we have

$$a_0(u,v)=\left(u,Tv\right)_H. \tag{1.3.17}$$

Here T is a linear operator defined in W_1 with the range of values lying in H.

As the conditions of Lemma 1.3.1 hold, each function $v(x)$ from the set W_1 may be considered as a solution of problem (1.3.6). Besides, this function $v(x)$ will belong to the space $W_2^1(D)$ and it will vanish in the set $\sigma_{00} \cup \sigma_0^{01} \cup \gamma_{00}$. Conversely, each function $w(x)$ from the set W uniquely determines a function $v(x)$ from the set W_1, as a solution of problem (1.3.6); this function will belong to the space $W_2^1(D)$ and it will vanish in the set $\sigma_{00} \cup \sigma_0^{01} \cup \gamma_{00}$.

Thus, if the functions $w(x)$ run though the set W, then the functions $v(x)$ run though all the set W_1.

Taking into account the facts that v belongs to the space $W_2^1(D)$, vanishes in the set $\sigma_{00} \cup \sigma_0^{01} \cup \gamma_{00}$, that the function $l_0 v$ vanishes in the set Γ_3, and the hypotheses of the theorem, we may easily show that

$$\left| a_0(v, v) \right| \geq \lambda_0 \left\| v \right\|_H^2, \qquad \lambda_0 > 0.$$

This inequality yields the following one

$$\left\| Tv \right\|_H \geq \lambda_0 \left\| v \right\|_H.$$

This inequality means that the operator T maps one-to-one the set W_1 onto the set $T(W_1)$.

We denote by H_1 the closure of the set $T(W_1)$ with respect to the norm of the space H.

For the functions $v(x)$ from the set W_1, we define the functional $\tilde{f}(v)$ as follows:

$$\tilde{f}(v) = \int_D f l_0 v dx.$$

We show now that this functional is a linear continuous functional on H_1. Indeed, we have

$$\left| \int_D f l_0 v dx \right| \leq \left\| f \right\|_{L_2(D)} \cdot \left\| l_0 v \right\|_{L_2(D)} \leq \left\| f \right\|_{L_2(D)} \cdot \left\| v \right\|_H$$

$$\leq \frac{1}{\lambda_0} \left\| f \right\|_{L_2(D)} \cdot \left\| Tv \right\|_H,$$

whence the required statement follows.

As before, by the Riesz theorem, there exists such an element u from the space H_1, that

$$\tilde{f}(v) = \left(u, Tv\right)_H$$

for each element v from the set W_1.

So, we have found the function $u(x)$ from the space H_1 (and, therefore, from the space H), which satisfies the identity

$$\int_D f l_0 v dx = \left(u, Tv\right)_H$$

for each function $v(x)$ from the set W_1. On the other hand, equality (1.3.17) means that this identity may be transformed as follows:

$$\int_D f l_0 v dx = a_0(u, v),$$

and, further, it may be transformed to the form

$$\int_D f l_0 v dx = a_0(u, l_0 v).$$

As the function $l_0 v$ is an arbitrary function from the set W, the last identity denotes that, for the function $u(x)$ from the space H, the following identity

$$a(u, w) = \int_D f w dx$$

will hold for each function $w(x)$ from the set W.

This function $u(x)$ will be the desired generalized solution of the boundary-value problem I. The theorem is proved.

It is easy to show that if a generalized solution $u(x)$ is a smooth function, then, under the hypotheses of the theorem, it will be a regular solution.

Conversely, each regular solution of the boundary-value problem I will be a generalized solution.

Remark. Here and further we shall not cite the results on the boundary-value problem I′. Each careful reader may obtain these results independently using the ideas stated in this book while obtaining the results on the boundary-value problem I.

1.3.3. Existence of regular solutions of the hyperbolic boundary-value problem

Description of all classes of composite type equations (1.3.1) for which boundary-value problem (1.3.2), (1.3.3) (or (1.3.2′), (1.3.3′)) has regular solutions is a problem which is as yet unsolved. In this book we shall describe only some such classes. In some cases we shall give the proofs; and some cases we shall give only the scheme of the proof.

The simplest class of equations (1.3.1) for which the hyperbolic boundary-value problem has a regular solution is the class of such equations with the elliptic operator A.

So, we suppose that the operator A in equation (1.3.1) is elliptic. It is known that for elliptic operators of second order the second basic inequality holds. Namely, *for functions $v(x)$ from the space $W_2^2(D)$ which vanish in the boundary of the domain D, we have the inequality*

$$\kappa_0 \left\| v \right\|^2_{W_2^2(D)} \le \left\| Av \right\|^2_{L_2(D)} + \kappa_1 \left\| v \right\|^2_{L_2(D)}$$

with a certain fixed positive constant κ_0 (Ladyzhenskaya and Ural'tseva, 1973).

In order that the second basic inequality hold it needs that the boundary satisfies certain conditions of smoothness. In our case it suffices that the boundary belongs to the class $B^{3,\beta}$.

Suppose now that the operator C is represented in the form

$$Cu = l_1 l_0 u + C^0 u, \tag{1.3.18}$$

where l_1 is an operator of first order; C^0 is a second order operator of the form

$$C^0 u = c^{ij,0}(x)u_{x_i x_j} + c^{i,0}(x)u_{x_i} + c^0(x)u.$$

Define the constant M_0 as follows:

$$M_0 = \max_{i,j}\left[\max_{\bar{D}}\left|c^{ij,0}(x)\right|.\right]$$

Besides the constants κ_0 and M_0, we shall need the constant R_1 – the constant from inequality (1.3.9) of Lemma 1.3.1.

Theorem 1.3.3. *Suppose that the operator A is elliptic, the operator C is symmetric and is of form (1.3.18) and let the following conditions hold*

$$d^{ij}\xi_i\xi_j + a_0(\alpha^i\xi_i)^2 \geq 0, \quad (\alpha^k c)_{x_k} \geq 0, \quad \textit{for } x \in \bar{D},\ \xi \in R^n;$$

$$c \geq 0, c^{ij}\xi_i\xi_j \geq 0 \quad \textit{for} \quad x \in \sigma_2;$$

$$\alpha^k \nu_k \geq \bar{\alpha}_0 > 0, \quad \textit{for} \quad x \in \sigma_1;$$

$$\kappa_0 - nM_0^2 R_1^2 > 0.$$

Let, besides, for the set σ_0, condition (1.3.7) hold. Then, for each function $f(x)$ from the space $L_2(D)$, there exists a solution $u(x)$ of the boundary-value problem I such that $u(x)$ attains boundary conditions (1.3.2) and (1.3.3); $u(x)$ and $l_0u(x)$ belong to the space $W_2^2(D)$.

Proof. First we note that the statement of the theorem denotes that equation (1.3.1) is replaced by the equation

$$Al_0u + Cu = f(x), \tag{1.3.1'}$$

and the existence of just this equation is considered in the theorem.

Consider now the set of equations

$$L_\tau u \equiv Al_0u + \tau Cu = f(x), \qquad \tau \in [0,1], \tag{1.3.1'_\tau}$$

and the set of boundary-value problems: *find solutions of equations (1.3.1$'_\tau$) satisfying the conditions*

$$u\Big|_{\sigma_1} = 0, \qquad l_0 u\Big|_{\Gamma} = 0. \tag{1.3.19}$$

Evidently, for $\tau = 0$, boundary-value problem (1.3.1$'_0$), (1.3.19) has a solution from the class which is required in the theorem. Really, the second of boundary conditions (1.3.19) allows us to obtain the function $v = l_0 u$ from the space $W_2^2(D)$ as the Dirichlet problem solution for an elliptic equation. Further, Lemma 1.3.1 allows us to determine the function u; and this function will belong to the space $W_2^2(D)$ and will vanish in the set σ_0.

We show now that for solutions of boundary-value problem (1.3.1$'_\tau$), (1.3.19), the estimates, uniform with respect to τ, hold

$$\Big|\Big|u\Big|\Big|_{W_2^1(D)} + \Big|\Big|l_0 u\Big|\Big|_{W_2^1(D)} \le M_1, \tag{1.3.20}$$

$$\Big|\Big|u\Big|\Big|_{W_2^2(D)} + \Big|\Big|l_0 u\Big|\Big|_{W_2^2(D)} \le M_2. \tag{1.3.21}$$

Here the constants M_1 and M_2 depend only on the function $f(x)$, the coefficients of the operator L, and on the domain D.

To establish the first estimate, we consider the equality

$$\int\limits_D L_\tau u \cdot l_0 u dx = \int\limits_D f l_0 u dx.$$

Integrating by parts, using the hypotheses of the theorem and the equalities $u_{x_i} = 0$ in σ_1 $(i = 1, \dots, n)$, we may obtain the estimate

$$\Big|\Big|l_0 u\Big|\Big|_{W_2^1(D)} \le M_1'$$

with the constant M_1' dependent only on the function $f(x)$ and on the coefficients of the operator L. Using estimate (1.3.8), we obtain the first estimate (1.3.20).

Consider now the equality

$$\int_D L_\tau u \cdot Al_0 u dx = \int_D f Al_0 u dx.$$

We may rewrite this equality as follows:

$$\int_D \left(Al_0 u\right)^2 dx = \int_D f Al_0 u dx - \int_D Al_0 u \cdot l_1 l_0 u dx$$

$$- \int_D Al_0 u \cdot c^{ij,0} u_{x_i x_j} dx - \int_D Al_0 u \left(c^{i,0} u_{x_i} + c^0 u\right) dx. \tag{1.3.22}$$

All the terms in the right side of this equality, except for the third term, are estimated from above by the Young inequality and estimate (1.3.20) by the value

$$\rho \int_D \left(Al_0 u\right)^2 dx + M\left(\rho, M_1', f\right),$$

where ρ is an arbitrary positive constant. The third term is estimated by the Young inequality and estimate (1.3.9) by the value

$$\frac{1}{2} \int_D \left(Al_0 u\right)^2 dx + \frac{n M_0^2 R_1^2}{2} \left|\left| l_0 u \right|\right|^2_{W_2^1(D)}.$$

Estimating now the left side of equality (1.3.22) by means of the second basic inequality for elliptic operators, taking into account the conditions connecting the constants κ_0, M_0, and R_1, estimate (1.3.20), and choosing the constant ρ sufficiently small, we obtain required estimate (1.3.21).

The estimates (1.3.20), (1.3.21), uniform with respect to τ and the statement on existence of solution of boundary-value problem $(1.3.1_0')$, (1.3.19) allow us to obtain the set of boundary-value problems $(1.3.1_\tau')$, (1.3.19) using the classical method of continuation by parameter. Following this method, we may prove the existence of such a solution $u^\tau(x)$ and that $u^\tau(x)$ belongs to the space $W_2^2(D)$; $l_0 u^\tau(x)$ belongs to the space $W_2^2(D)$; and $u^\tau(x)$ attains boundary conditions (1.3.19) for all τ from the interval $[0, 1]$. For $\tau = 1$, this function will solve the equation $(1.3.1')$; by Lemma 1.3.1, this function will vanish in the set σ_0; therefore, it will be the desired solution. The theorem is proved.

As we have mentioned in §1, the theory of such equations with a singled out variable plays an important role in the theory of composite type equations. We shall give special attention to such equations in our book.

First we note that if the operator A is parabolic with the time variable x_1, and the domain D is a cylinder with the element parallel to the axis Qx_1, then the boundary-value problem I, undoubtedly, will have a regular solution. This may be proved analogously to the proof of Theorem 1.3.3.

Now, we shall give an interesting example of equations with a singled out variable which have regular solutions.

We suppose that Q is a cylindrical domain of the variables x and t:

$$Q = D \times (0,T), \quad x \in D, \quad t \in (0,T), \quad 0 < T < +\infty.$$

In this cylinder, we consider the equation

$$Lu \equiv K(t)u_{tt} - Au_t - Bu = f(x,t), \tag{1.3.23}$$

where A and B are elliptic-parabolic operators of the form

$$Au \equiv \frac{\partial}{\partial x_i}\left(a^{ij}(x)u_{x_j}\right) - a_0(x)u,$$

$$Bu \equiv \frac{\partial}{\partial x_i}\left(b^{ij}(x)u_{x_j}\right) - b_0(x)u.$$

The smooth function $K(t)$ satisfies the conditions $K(0) \geq 0$, $K(T) \geq 0$; the functions $a^{ij}(x)$, $b^{ij}(x)$, $a_0(x)$, and $b_0(x)$ are smooth functions in $\bar{D}$.

Evidently, equation (1.3.23) may be represented in form (1.3.1), but, in this case, the operator $K(t)u_t - Au$ will be a parabolic operator of a new class of degenerate equations of the composite type which has not been investigated before.

Note that if the operator A is elliptic and $K(t) \equiv 1$, or $K(t) \equiv 0$ and $a_0(x) \equiv 1$, then equations (1.3.23) are well-studied pseudohyperbolic or pseudoparabolic equations (see §1 of this chapter). In our case, the sign of $K(t)$ is not fixed; therefore, equations (1.3.23) may be either degenerate pseudoparabolic equations, or, in a certain subset of the cylinder Q, they may be of the quasielliptic type.

Equations (1.3.23) the author will consider as the model equations. On the basis of these equations, the methods of establishing the existence of regular solutions will be clarified. Of course, these methods may be applied in other situations.

We suppose, for definiteness, that in equation (1.3.23) the condition $K(0) = 0$ holds. We may use the notations of point 3.1 for setting of the boundary-value problem. But, in order to underline that the ideas stated below are applicable in other situations, we introduce independent notations. Namely, we denote

$$\sigma_0(A) = \Big\{x \in \Gamma : a^{ij}\nu_i\nu_j = 0\Big\}, \quad \sigma_1(A) = \Big\{x \in \Gamma : a^{ij}\nu_i\nu_j > 0\Big\},$$

$$\sigma_0(B) = \Big\{x \in \Gamma : b^{ij}\nu_i\nu_j = 0\Big\}, \quad \sigma_1(B) = \Big\{x \in \Gamma : b^{ij}\nu_i\nu_j > 0\Big\}.$$

According to the setting of the boundary-value problem I, we consider the following problem: *find a solution of equation (1.3.23) which satisfies the conditions*

$$u(x,0) = 0, \qquad u(x,t)\Big|_{[\sigma_1(A)\cup\sigma_1(B)]\times(0,T)} = 0. \tag{1.3.24}$$

We suppose that the elliptic-parabolic condition for operator A and B holds in the following form: *there exist infinitely differentiable in $\bar{D}$ functions $a^{i,0}(x)$ and $b^{i,0}(x)$ such that*

$$0 \le a^{i,0}\xi_i^2 \le a^{ij}\xi_i\xi_j \le \lambda_1 a^{i,0}\xi_i^2,$$

$$0 \le b^{i,0}\xi_i^2 \le b^{ij}\xi_i\xi_j \le \lambda_2 b^{i,0}\xi_i^2, \tag{1.3.25}$$

where $\lambda_1 > 0$, $\lambda_2 > 0$, $x \in \bar{D}$, $\xi \in R^n$.

We define the condition of **characteristic convexity** which we shall need not only in this section.

Suppose that M and R are symmetric elliptic-parabolic operators of the form

$$Mu = \frac{\partial}{\partial x_i}\Big(m^{ij}(x)u_{x_j}\Big), \qquad Ru = \frac{\partial}{\partial x_i}\Big(r^{ij}(x)u_{x_j}\Big).$$

We denote by Γ^0_{MR} the set of points from Γ, where the operators M and R are elliptic simultaneously. For a (sufficiently small) positive number δ, we denote by $\Gamma^{0,\delta}_{MR}$ the set of points of Γ^0_{MR} outstanding from the boundary of Γ^0_{MR} at a distance greater than δ. By $\Gamma^{1,\delta}_{MR}$ we denote the set $\Gamma\backslash\Gamma^{0,\delta}_{MR}$. As the boundary Γ of the domain D is compact and piecewise smooth, it may be represented as a union of a finite number of connected components $\Gamma_1,\dots,\Gamma_m$ of dimension $n-1$ and of a set Γ_{m+1} of dimension less than $n-1$. In this case, each component Γ_s for $s \leq m$, is described by the equation $F_s(x) = 0$, where $F_s(x)$ is a smooth function.

Without loss of generality (using the refinement and enumerating if necessary), we may consider that the set $\Gamma^{1,\delta}_{MR}$ is formed by the components $\Gamma_1,\dots,\Gamma_{m_0}$ ($m_0 \leq m$) and by the subset Γ^1_{m+1} of the set Γ_{m+1}; and the set $\Gamma^{0,\delta}_{MR}$ is formed by the remaining components of the dimension $n-1$ and the set $\Gamma_{m+1}\backslash\Gamma^1_{m+1}$. We shall make on the component Γ_s the transformation

$$y = \mathcal{F}_s(x),$$

or in coordinate form $y_k = F^k_s(x)$ which transforms the indicated component into the part of the plane $y_n = 0$ (for the transformation $\mathcal{F}_s$ we have $F^n_s \equiv F_s$). Note that the transformation $\mathcal{F}_s$ is reversible. Further, if in the surface integral

$$\int\limits_{\Gamma_s} h(x)d\gamma,$$

we pass to the $(n-1)$-dimensional integral over the part of the plane $y_n = 0$, then this integral becomes as follows:

$$\int\limits_{\mathcal{F}_s(\Gamma_s)} h\psi_s dy_1 \dots dy_{n-1},$$

where $\psi_s(y)$ is the modulus of the transition from the surface integral to the integral over the plane.

We define now *the (M,R)-condition, or the condition of characteristic convexity of the domain D relative to the operators M and R: in the sets $\mathcal{F}_s(\Gamma_s)$, for $s = 1,\dots,m_0$, the following condition holds:*

$$m^{ij}r^{kl}F^n_{sx_j}\left(F^n_{sx_ix_l}\nu_k - F^n_{sx_kx_l}\nu_i\right)\psi_s + m^{ij}F^n_{sx_j}F^n_{sx_l}\left(r^{kl}_{x_i}\nu_k - r^{kl}_{x_l}\nu_i\right)\psi_s$$

$$-\frac{1}{2}\sum_{q<n}\left[m^{ij}r^{kl}F^n_{sx_j}\left(F^q_{sx_i}\nu_k - F^q_{sx_k}\nu_i\right)\psi_s\right]_{y_q} \geq 0.$$

Suppose that the operators A_0 and B_0 are as follows:

$$A_0u = \frac{\partial}{\partial x_i}\left(a^{ij}(x)u_{x_j}\right), \qquad B_0u = \frac{\partial}{\partial x_i}\left(b^{ij}(x)u_{x_j}\right).$$

We define, also, the constants α_0, β_0, and κ_0:

$$\alpha_0 = \max_{\bar{D}}\left[\sum_{i=1}^{n} a^{i,0}(x)\right], \qquad \beta_0 = \max_{\bar{D}}\left[\sum_{i=1}^{n} b^{i,0}(x)\right],$$

$$\kappa_0 = \max_{i,j}\max_{\bar{D}}\left[\sum_{i=1}^{n} a^{il}_{x_k}(x)b^{kj}_{xl}(x)\right].$$

Theorem 1.3.4. *Let operators A and B be symmetric, $A+B$ be elliptic in D with the ellipticity constant μ_0, and suppose that the conditions (1.3.25), (A_0, B_0)-condition, and the conditions*

$$\left|a^{ij}_{x_k}(x)\right| \leq M_0\sqrt{a^{i,0}(x)}, \qquad \left|b^{ij}_{x_k}(x)\right| \leq M_0\sqrt{b^{i,0}(x)},$$

where

$$x \in \bar{D}, \quad i,j,k = 1,\dots,n, \quad M_0 = \text{const} \geq 0;$$

$$b_0(x) \geq 0, \quad a_0(x) \geq \bar{a}_0 > 0 \quad \text{for} \quad x \in \bar{D},$$

hold. Here we choose the constant $\bar{a}_0$ so large that for all t from the interval $[0,T]$ and the number $\lambda_0 > \frac{1}{2}$ the inequalities

$$\bar{a}_0 - \frac{1}{2}K_t(t) - \frac{\lambda_0}{2}K(t) \geq \bar{a}_1 > 0,$$

$$\bar{a}_0\mu_0 - \frac{1}{2}K_t(t) + \frac{\lambda_0}{2}K(t) - \kappa_0 - M_0^2 n^2\Big(\lambda_1^2\alpha_0 + \lambda_2^2\beta_0\Big) \geq \bar{a}_1 > 0$$

hold. Then for each function $f(x,t)$ vanishing for $x \in D$, $t = 0$ and such that

$$\int\limits_Q \Big[f^2 + f_t^2 + b^{ij} f_{x_i t} f_{x_j t}\Big] dxdt < +\infty,$$

there exists a solution $u(x,t)$ of problem (1.3.23), (1.3.24) such that

$$u \in L_2\Big(0,T; W_2^2(D) \bigcap \overset{0}{W_2^1}(D)\Big),$$

$$u_t \in L_2\Big(0,T; W_2^2(D) \bigcap \overset{0}{W_2^1}(D)\Big),$$

$$u_{tt} \in L_2(Q).$$

Proof. For $\varepsilon > 0$, we consider the set of operators

$$L_\varepsilon u \equiv -\varepsilon u_{ttt} + K u_{tt} - A_{0,\varepsilon} u_t + a_0 u_t - Bu,$$

where $A_{0,\varepsilon} = A_0 + \varepsilon B_0$ and the set of the corresponding boundary-value problems: *find a solution of the equation*

$$L_\varepsilon u = f(x,t), \tag{1.3.23$_\varepsilon$}$$

which satisfies conditions (1.3.24) and the conditions

$$u_t(x,0) = 0, \qquad u_{tt}(x,T) = 0. \tag{1.3.24$_\varepsilon$}$$

We shall establish now some *a priori* estimates of the regular solutions of problem (1.3.23$_\varepsilon$), (1.3.24), (1.3.24$_\varepsilon$). These estimates will enable us both to establish the solvability of the regularized problem and to pass to the limit for $\varepsilon \to 0$ and, thus, to obtain the required solution of problem (1.3.23), (1.3.24).

First of all, we note that the equality

$$\int\limits_Q L_\varepsilon u \cdot u_t \exp\Big(-\lambda_0 t\Big)dxdt = \int\limits_Q f u_t \exp\Big(-\lambda_0 t\Big)dxdt$$

allows us to obtain the first *a priori* estimate

$$\varepsilon \int\limits_Q u_{tt}^2 dxdt + \int\limits_Q \Big[u_t^2 + a^{ij}u_{x_i t}u_{x_j t} + b^{ij}u_{x_i}u_{x_j}\Big]dxdt \le N_1, \qquad (1.3.26)$$

where the constant N_1 depends only on the coefficients of the operator L and on the function f.

The equality

$$\int\limits_Q L_\varepsilon u \cdot u_{ttt} \exp\Big(-\lambda_0 t\Big)dxdt = \int\limits_Q f u_{ttt} \exp\Big(-\lambda_0 t\Big)dxdt,$$

after integrating by parts (in both sides), using the theorem conditions and estimate (1.3.26) allows us to obtain the second *a priori* estimate

$$\varepsilon \int\limits_Q u_{ttt}^2 dxdt + \int\limits_Q \Big[u_{tt}^2 + a^{ij}u_{x_i tt}u_{x_j tt} + b^{ij}u_{x_i t}u_{x_j t}\Big]dxdt \le N_2, \qquad (1.3.27)$$

where the constant N_2 depends on the coefficients of the operator L and on the function f also.

The analysis of the equality

$$\int\limits_Q L_\varepsilon u \frac{\partial}{\partial t}\Big(\exp\Big(-\lambda_0 t\Big)(A_0 + B_0)u_{tt}\Big)dxdt$$

$$= \int\limits_Q f \frac{\partial}{\partial t}\Big(\exp\Big(-\lambda_0 t\Big)(A_0 + B_0)u_{tt}\Big)dxdt \qquad (1.3.28)$$

will complete the establishment of the required estimates.

Firstly, the corollary of equality (1.3.28) is the equality

$$\varepsilon \int\limits_Q \Big(a^{ij}+b^{ij}\Big)u_{x_i ttt}u_{x_j ttt}\exp\Big(-\lambda_0 t\Big)dxdt$$

$$+\frac{\varepsilon\lambda_0}{2}\int\limits_Q \Big(a^{ij}+b^{ij}\Big)u_{x_i tt}(x,0)u_{x_j tt}(x,0)dx$$

$$+\varepsilon\int\limits_Q \Big(B_0u_{tt}\Big)^2\exp\Big(-\lambda_0 t\Big)dxdt+\int\limits_Q \Big(A_0u_{tt}\Big)^2\exp\Big(-\lambda_0 t\Big)dxdt$$

$$+\frac{\lambda_0}{2}\int\limits_Q \Big(B_0u_t\Big)^2\exp\Big(-\lambda_0 t\Big)dxdt+\frac{1}{2}\exp\Big(-\lambda_0 T\Big)\int\limits_D \Big[B_0u_t(x,T)\Big]^2dx$$

$$+\int\limits_Q \Big(a_0+\frac{1}{2}K_t+\frac{\lambda_0}{2}K\Big)\Big(a^{ij}+b^{ij}\Big)u_{x_i tt}u_{x_j tt}\exp\Big(-\lambda_0 t\Big)dxdt$$

$$+(1+\varepsilon)\int\limits_Q A_0u_{tt}\cdot B_0u_{tt}\exp\Big(-\lambda_0 t\Big)dxdt=-\int\limits_Q A_0u_{tt}\cdot B_0u_t\exp\Big(-\lambda_0 t\Big)dxdt$$

$$-\int\limits_Q a_{0x_i}\Big(a^{ij}+b^{ij}\Big)u_{x_j tt}u_{tt}\exp\Big(-\lambda_0 t\Big)dxdt$$

$$-\int\limits_Q b_0\Big(a^{ij}+b^{ij}\Big)u_{x_i t}u_{x_j tt}\exp\Big(-\lambda_0 t\Big)dxdt$$

$$-\int\limits_Q b_{0x_i}\Big(a^{ij}+b^{ij}\Big)u_{x_j tt}u_t\exp\Big(-\lambda_0 t\Big)dxdt-\int\limits_Q f_tA_0u_{tt}\exp\Big(-\lambda_0 t\Big)dxdt$$

$$+\int\limits_Q f_{x_i t}b^{ij}u_{x_j tt}\exp\Big(-\lambda_0 t\Big)dxdt. \qquad (1.3.29)$$

We transform the last integral in the left side of equality (1.3.29). As all the calculations below will relate to the variables $x_1, \dots, x_n$, we shall transform, first, the integral

$$\int\limits_D A_0 u_{tt} \cdot B_0 u_{tt} dx.$$

After applying twice the integration by parts, we obtain the equality

$$\int\limits_D A_0 u_{tt} \cdot B_0 u_{tt} dx = \int\limits_D a^{ij} b^{kl} u_{x_i x_k tt} u_{x_j x_l tt} dx + \int\limits_D a^{ij}_{x_k} b^{kl} u_{x_i x_l tt} u_{x_j tt} dx$$

$$+ \int\limits_D a^{ij} b^{kl}_{x_i} u_{x_j x_k tt} u_{x_l tt} dx + \int\limits_D a^{ij}_{x_k} b^{kl}_{x_i} u_{x_j tt} u_{x_l tt} dx$$

$$+ \int\limits_\Gamma \Big[a^{ij} b^{kl} u_{x_j tt} u_{x_i x_l tt} \nu_k + a^{ij} b^{kl}_{x_i} u_{x_j tt} u_{x_l tt} \nu_k$$

$$- a^{ij} b^{kl} u_{x_j tt} u_{x_k x_l tt} \nu_i - a^{ij} b^{kl}_{x_k} u_{x_j tt} u_{x_l tt} \nu_i \Big] d\gamma. \qquad (1.3.30)$$

The first term in the right side of this equality is estimated from below by the integral

$$I_0 = \int\limits_D a^{i,0} b^{j,0} u^2_{x_i x_j tt} dx.$$

To estimate the remain volume integrals

$$I_1 = \int\limits_D a^{ij}_{x_k} b^{kl} u_{x_i x_l tt} u_{x_j tt} dx, \quad I_2 = \int\limits_D a^{ij} b^{kl}_{x_i} u_{x_j x_k tt} u_{x_l tt} dx,$$

$$I_3 = \int\limits_D a^{ij}_{x_k} b^{kl}_{x_i} u_{x_j tt} u_{x_l tt} dx,$$

we estimate their integrand expressions.

In order to estimate the integrand expression of the integral I_1, we set

$$\xi_l^i = u_{x_i x_l tt}, \qquad \mu_k^i = \sum_{j=1}^{n} a_{x_k}^{ij} u_{x_j tt}$$

(here and after, in order to clarify the calculations, we shall indicate the summation by the sign $\sum$). The Cauchy-Schwarz-Bunyakovskii inequality and the hypotheses of the theorem yield the following inequalities

$$\left| \sum_{k,l=1}^{n} b^{kl} \xi_l^i \mu_k^i \right| \leq \sqrt{\sum_{k,l=1}^{n} b^{kl} \xi_k^i \xi_l^i} \cdot \sqrt{\sum_{k,l=1}^{n} b^{kl} \mu_k^i \mu_l^i}$$

$$\leq \lambda_2 \sqrt{\sum_{k=1}^{n} b^{k,0} (\xi_k^i)^2} \cdot \sqrt{\sum_{k=1}^{n} b^{l,0} (\mu_l^i)^2},$$

$$\sum_{l=1}^{n} b^{l,0} (\mu_l^i)^2 \leq \sum_{l=1}^{n} b^{l,0} \left| \sum_{j=1}^{n} a_{x_k}^{ij} u_{x_j tt} \right|^2 \leq \sum_{l=1}^{n} b^{l,0} \left(\sum_{j=1}^{n} \left| a_{x_l}^{ij} \right| \left| u_{x_j tt} \right| \right)^2$$

$$\leq M_0^2 a^{i,0} \sum_{l=1}^{n} b^{l,0} \left(\sum_{j=1}^{n} \left| a_{x_j tt} \right| \right)^2 \leq M_0^2 \beta_0 n a^{i,0} \sum_{j=1}^{n} u_{x_j tt}^2.$$

Further, we have

$$\left| a_{x_k}^{ij} b^{kl} u_{x_i x_l tt} u_{x_j tt} \right| = \left| \sum_{i=1}^{n} \sum_{k,l=1}^{n} b^{kl} \xi_l^i \mu_k^i \right|$$

$$\leq \sum_{i=1}^{n} \left| \sum_{k,l=1}^{n} b^{kl} \xi_l^i \mu_k^i \right| \leq \sum_{i=1}^{n} \left[\lambda_2 \sqrt{M_0^2 \beta_0 n} \cdot \sqrt{a^{i,0} b^{k,0} u_{x_i x_k tt}^2} \cdot \sqrt{\sum_{j=1}^{n} u_{x_j tt}^2} \right]$$

$$\leq \frac{\rho^2}{2} a^{i,0} b^{k,0} u_{x_i x_k tt}^2 + \frac{\lambda_2^2 M_0^2 \beta_0 n^2}{2\rho^2} \sum_{j=1}^{n} u_{x_j tt}^2$$

(the last inequality is obtained taking account of the Young inequality; ρ is an arbitrary positive number).

Hence we obtain the estimate for the integral I_1

$$|I_1| \leq \frac{\rho^2}{2} I_0 + \frac{\lambda_2^2 M_0^2 \beta_0 n^2}{2\rho^2} \Big|\Big| u_{tt} \Big|\Big|^2_{L_2(0,T;\overset{\circ}{W}{}_2^1(D))}.$$

Similarly, we may obtain the estimate for the integral I_2

$$|I_2| \leq \frac{\rho^2}{2} I_0 + \frac{\lambda_1^2 M_0^2 \alpha_0 n^2}{2\rho^2} \Big|\Big| u_{tt} \Big|\Big|^2_{L_2(0,T;\overset{\circ}{W}{}_2^1(D))}.$$

Finally, the integral I_3 is estimated as follows

$$|I_3| \leq \kappa \Big|\Big| u_{tt} \Big|\Big|_{L_2(0,T;\overset{\circ}{W}{}_2^1(D))}.$$

Now, we fix the number ρ setting $\rho^2 = 1/2$. Then the volume integrals in the right side of equality (1.3.30) are estimated from below by the value

$$\frac{1}{2} I_0 - \Big[(\lambda_1^2 \alpha_0 + \lambda_2^2 \beta_0) M_0^2 n^2 + \kappa_0 \Big] \Big|\Big| u_{tt} \Big|\Big|^2_{L_2(0,T;\overset{\circ}{W}{}_2^1(D))}.$$

We consider now the boundary integral from the right side of equality (1.3.30).

We represent this integral as the sum of the integrals over the sets $\Gamma_1, \ldots, \Gamma_m$, and each of these integrals we transform to the plane integral. As a result, we obtain

$$I_\Gamma = \int\limits_\Gamma \Big[a^{ij} b^{kl} \Big(u_{x_j tt} u_{x_i x_l tt} \nu_k - u_{x_j tt} u_{x_k x_l tt} \nu_i \Big)$$

$$+ \Big(a^{ij} b^{kl}_{x_i} u_{x_j tt} u_{x_l tt} \nu_k - a^{ij} b^{kl}_{x_k} u_{x_j tt} u_{x_l tt} \nu_i \Big) \Big] d\gamma$$

$$= \sum_{s=1}^{m} \int\limits_{\mathcal{F}_s(\Gamma_s)} \Big\{ a^{ij} b^{kl} \Big[\Big(F^p_{s x_j} u_{y_p tt} \Big) \Big(F^p_{s x_i} F^q_{s x_l} u_{y_p y_q tt} + F^p_{s x_i x_l} u_{y_p tt} \Big) \nu_k$$

$$- \Big(F^p_{s x_j} u_{y_p tt} \Big) \Big(F^p_{s x_k} F^q_{s x_l} u_{y_p y_q tt} + F^p_{s x_k x_l} u_{y_p tt} \Big) \nu_i \Big] \psi_s \Big\} dy_1 \ldots dy_{n-1}$$

$$+\sum_{s=1}^{m}\int\limits_{\mathcal{F}_s(\Gamma_s)}\Big[a^{ij}b^{kl}_{x_i}\Big(F^p_{sx_j}u_{y_ptt}\Big)\Big(F^q_{sx_l}u_{y_qtt}\Big)\nu_k$$

$$-a^{ij}b^{kl}_{x_i}\Big(F^p_{sx_j}u_{y_ptt}\Big)\Big(F^q_{sx_l}u_{y_qtt}\Big)\nu_i\Big]\psi_s dy_1\ldots dy_{n-1}.$$

The equalities $u_{y_itt}=u_{y_iy_jtt}=0$ for $i,j=1,\ldots,n-1$ hold. As the component Γ_s is described by the equation $F^n_s=0$, then the equalities

$$F^n_{sx_i}\nu_k=F^n_{sx_k}\nu_i;\qquad i,k=1,\ldots,n$$

also hold. Taking them into account, we transform the boundary integral I_Γ as follows:

$$I_\Gamma=\sum_{s=1}^{m}\int\limits_{\mathcal{F}_s(\Gamma_s)}\sum_{q<n}\Big[a^{ij}b^{kl}F^n_{sx_j}\Big(F^q_{sx_i}\nu_k-F^q_{sx_k}\nu_i\Big)\psi_s\Big]u_{y_ntt}u_{y_ny_qtt}dy_1\ldots dy_{n-1}$$

$$+\sum_{s=1}^{m}\int\limits_{\mathcal{F}_s(\Gamma_s)}\Big[a^{ij}b^{kl}F^n_{sx_j}\Big(F^n_{sx_ix_l}\nu_k-F^n_{sx_kx_l}\nu_i\Big)\psi_s+a^{ij}F^n_{sx_j}F^n_{sx_l}$$

$$\times\Big(b^{kl}_{x_i}\nu_k-b^{kl}_{x_k}\nu_i\Big)\psi_s\Big]u^2_{y_ntt}dy_1\ldots dy_{n-1}. \tag{1.3.31}$$

In each integral from the first sum we integrate by parts with respect to the variable y_q. Then, these integrals will be transformed to the integrals which contain only the derivative u_{y_ntt}. The integrals over the boundaries of the sets $\mathcal{F}(\Gamma_s)$ will be mutually cancelling. In the obtained sum we select the terms corresponding to the indices $1,\ldots,m_0$. By the (A_0,B_0)–condition, these terms will be nonnegative. Each of the integrals over the sets $\mathcal{F}_s(\Gamma_s)$, $s=m_0+1,\ldots,m$ may be estimated from above by the value

$$N_3\int\limits_{\mathcal{F}_s(\Gamma_s)}u^2_{y_ntt}dy_1\ldots dy_{n-1},$$

where the constant N_3 depends only on the coefficients of the operator L and the domain D. We return in the last integral to the variables $x_1, \dots, x_n$. Then, after summation, it may be shown easily that the sum of all the terms in the right side of equality (1.3.31) corresponding to the indices $m_0 + 1, \dots, m$ is estimated from above by the value

$$N_4 \sum_{i=1}^{n} \left|\left| u_{x_i tt} \right|\right|^2_{L_2(\Gamma^{0,\sigma}_{A_0 B_0})}$$

with the constant N_4 dependent only on the coefficients of the operator L and the domain D. Note that, in the set $\Gamma^{0,\sigma}_{A_0 B_0}$, both operators A_0 and B_0 are uniformly elliptic. Therefore, there exists a subdomain D^δ of the domain D for which the set $\Gamma^{0,\sigma}_{A_0 B_0}$ is a part of the boundary, where both operators A_0 and B_0 are uniformly elliptic and whose boundary is piecewise smooth. In the domain D^δ, the inequality

$$\sum_{i=1}^{n} \int\limits_{\Gamma^{0,\sigma}_{A_0 B_0}} u^2_{x_i tt} d\gamma \le \rho_1 \sum_{i,j=1}^{n} \int\limits_{D^\delta} u^2_{x_i x_j tt} dx + N(\rho_1) \int\limits_{D^\delta} u^2_{tt} dx \qquad (1.3.32)$$

holds, where ρ_1 is an arbitrary positive constant (Sobolev, 1950; Ladyzhenskaya and Ural'tseva, 1973). Further, in the domain D^δ, the functions $a^{i,0}$ and $b^{j,0}$ are positive and are separated from zero by a certain positive constant k_0. Inequality (1.3.32) and estimate (1.3.27) yield the inequalities

$$N_4 \sum_{i=1}^{n} \int\limits_{\Gamma^{0,\sigma}_{A_0 B_0}} u^2_{x_i tt} d\gamma \le N_4 \rho_1 \sum_{i,j=1}^{n} \int\limits_{D^\delta} u^2_{x_i x_j tt} dx + N_s N(\rho_1)$$

$$\le \frac{N_4 \rho_1}{k_0^2} \int\limits_{D^\delta} a^{i,0} b^{j,0} u^2_{x_i x_j tt} dx + N_s N(\rho_1) \le \frac{N_4 \rho_1}{k_0^2} I_0 + N_5 N(\rho_1),$$

where the constant N_5 depends only on the coefficients of the operator L and the function f.

As a result of all these calculations, we obtain the inequality

$$\int\limits_{D} A_0 u_{tt} B_0 u_{tt} dx \ge \left(\frac{1}{2} - \frac{N_4 \rho_1}{h_0^2} \right) I_0$$

$$-\left[\left(\lambda_1^2\alpha_0+\lambda_2^2\beta_0\right)M_0^2n^2+\kappa_0\right]\left|\left|u_{tt}\right|\right|^2_{L_2(0,T;\overset{0}{W}{}_2^1(D))}-N_5N(\rho_1). \qquad (1.3.33)$$

Now, we consider again inequality (1.3.29). Taking into account inequality (1.3.33), estimate (1.3.27), the hypotheses of the theorem, and the Young inequality with the apposite constants, we obtain, finally, the estimate

$$\int_Q\left[\left(A_0u_{tt}\right)^2+\left(B_0u_t\right)^2\right]dxdt\le N_6, \qquad (1.3.34)$$

where the constant N_6 depends only the coefficients of the operator L, the function f, and the domain D.

Note that estimate (1.3.34) yields the estimate

$$\int_Q\left(A_0u_t\right)^2dxdt\le N_7$$

with the constant N_7 dependent on the constant N_6 and the number T. Hence follows the estimate

$$\int_Q\left[\left(A_0+B_0\right)u_t\right]^2dxdt\le N_8$$

with the constant N_8 dependent only on the constants N_6 and N_7. This estimate, estimate (1.3.26), and the second basic inequality for elliptic operators yield the estimate

$$\left|\left|u_t\right|\right|_{L_2(0,T;W_2^2(D))}\le N_0, \qquad (1.3.35)$$

where the constant N_0 depends only on the coefficients of the operator L, the function f, the domain D, and the number T. Finally, the uniform estimate of the function $u(x,t)$ in the space $L_2(0,T;W_2^2(D))$ is evident.

Now we shall pass directly to the proof of the theorem.

First, we note that, for ε fixed, problem ($1.3.23_\varepsilon$), (1.3.24), ($1.3.24_\varepsilon$) has a regular solution. Really, the operator $A_{0,\varepsilon}$ is an elliptic operator; therefore, for the equation

$$-\varepsilon u_{ttt} - \Big(A_{0,\varepsilon} - a_0\Big) u_t = f(x,t), \tag{1.3.36}$$

the boundary-value problem with conditions (1.3.24) and (1.3.24$_\varepsilon$) has a regular solution. Solvability of problem (1.3.36), (1.3.24), (1.3.24$_\varepsilon$) may be shown, for example, using the Fourier method. We consider now the set of equations

$$-\varepsilon u_{ttt} - \Big(A_{0,\varepsilon} - a_0\Big) u_t + \tau\Big(K u_{tt} - Bu\Big) = f(x,t), \quad \tau \in [0,1]. \tag{1.3.37}$$

For solutions of these equations, which satisfy conditions (1.3.24) and (1.3.24$_\varepsilon$), the following estimate

$$\int\limits_Q \left[u_{ttt}^2 + u_{tt}^2 + \sum_{i,j=1}^{n} \Big(u_{x_i x_j t}^2 + u_{x_i x_j}^2 \Big) \right] dxdt \le N \tag{1.3.38}$$

holds with the constant N dependent on ε, the coefficients of the operator L, the domain D, the number T, and the norm of f in the space $L_2(Q)$. The proof of estimate (1.3.38) is based on the analysis of the equalities which have led to estimates (1.3.26) and (1.3.27) and on using the second inequality for elliptic operators. Therefore, we may apply the method of continuation in a parameter to the set of boundary-value problems (1.3.37), (1.3.24), (1.3.24$_\varepsilon$). This method gives that boundary-value problem (1.3.23$_\varepsilon$), (1.3.24), (1.3.24$_\varepsilon$) has a regular solution for fixed ε.

We suppose that $u^\varepsilon(x,t)$ is a solution of problem (1.3.23$_\varepsilon$), (1.3.24), (1.3.24$_\varepsilon$). For the set of functions $\{u^\varepsilon(x,t)\}$, the estimates (1.3.26), (1.3.27), and (1.3.35), uniform with respect to ε, will hold. These estimates allow us to pass to the limit for $\varepsilon \to 0$ and to obtain the required solution. The theorem is proved.

Remark 1. The solution $u(x,t)$ of problem (1.3.23), (1.3.24) will be more smooth than the theorem states. In particular, the derivative u_{tt} belongs to the space $L_2(0,T; W_2^1(D))$.

Remark 2. If, for the function $K(t)$, we have $K(0) > 0$, then the theorem holds analogous to Theorem 1.3.4 for the boundary-value problem:

find a solution of equation (1.3.23) satisfying the conditions (1.3.24) and the condition

$$u_t(x,0) = 0.$$

Remark 3. Dependence of the functions a^{ij}, b^{ij}, a_0, and b_0 on the variable t leads to not vital but rather tedious conditions.

Remark 4. Equations (1.3.23) in the case of a zero-order operator A ($a_{ij} = 0$) and an elliptic operator B were stated by Vragov (1977, 1983).

The condition of characteristic convexity first appeared in the author's papers and was frequently used for proving the existence of regular solutions in different classes of equations of the composite type (Kozhanov, 1992, 1994a, 1994b, 1997a, 1997b). In the case when the operators A_0 and B_0 are not degenerate in the boundary of D, then the (A_0, B_0)-condition holds if the domain D is convex (Ladyzhenskaya and Ural'tseva, 1973). Evidently, the (A_0, B_0)-condition holds if the operator A_0 is degenerated to the zero order operator in a certain subset Γ_0 of the boundary of the domain D, and the operator B_0 is degenerated to the zero order operator in the complement of the set Γ_0. Finally, it is evident that the (A_0, B_0)-condition may be replaced by the (B_0, A_0)-condition.

1.3.4. The elliptic boundary-value problem. Uniqueness of regular solutions

We suppose that in equation (1.3.1) the operator A is an elliptic-parabolic operator, as before. We define the operator C_0 as follows:

$$C_0 u = \left(b^{ij} + \alpha a^{ij} - \alpha^k_{x_k} a^{ij}\right) u_{x_i x_j} + \left(b^i + \alpha a^i - \alpha^k_{x_k} a^i\right) u_{x_i}$$

$$+ \left(b + \alpha a - \alpha^k_{x_k} a\right) u \equiv c_0^{ij} u_{x_i x_j} + c_0^i u_{x_j} + c_0 u.$$

We suppose that the operator C_0 is elliptic-parabolic in the set σ_0^0. Select from this set the subsets s^0, s_1^0, s_2^0, s_0^0, and s_3

$$s^0 = \left\{x \in \sigma_0^0 : \; c_0^{ij} \nu_i \nu_j = 0\right\},$$

$$s_1^0 = \left\{x \in s^0 : \; \left(c^{ij}_{0x_j} - c_0^i\right) \nu_i > 0\right\},$$

$$s_2^0 = \left\{x \in s^0 : \ \left(c_{0x_j}^{ij} - c_0^i\right)\nu_i < 0\right\},$$

$$s_0^0 = \left\{x \in s^0 : \ \left(c_{0x_j}^{ij} - c_0^i\right)\nu_i = 0\right\},$$

$$s_3 = \left\{x \in \sigma_0^0 : \ c_0^{ij}\nu_i\nu_j > 0\right\}.$$

Boundary-value problem II: *Find a solution of equation (1.3.1) satisfying the conditions*

$$u\Big|_{\sigma_1\cup\sigma_2\cup\sigma_{00}\cup s_1^0\cup s_1} = 0, \tag{1.3.39}$$

$$l_0 u\Big|_{\sigma_{10}} = 0. \tag{1.3.40}$$

The boundary-value problem II, like the boundary-value problem I, generalizes the first boundary-value problem for elliptic-parabolic equations. Really, if we set $\alpha^k \equiv 0$, $\alpha \equiv 1$, $B \equiv 0$, then the operator C_0 will coincide with the operator A; boundary condition (1.3.40) will vanish; boundary condition (1.3.39) will transform into the boundary condition (Fichera, 1960; Oleinik and Radkevich, 1969) of the first boundary-value problem for elliptic-parabolic equations.

We define now the functions $d^{ij,0}(x)$

$$d_0^{ij} = c_0^{ij} - \left(\alpha^i a^{kj}\right)_{x_k} + \alpha^i a^j + \frac{1}{2}\left(\alpha^k a^{ij}\right)_{x_k}.$$

Theorem 1.3.5. *Let an operator A be symmetric and elliptic-parabolic in the set $\bar{D}$, an operator C_0 be elliptic-parabolic in the set σ_0^0 and suppose that the condition*

$$d_0^{ij}\xi_i\xi_j \geq 0 \quad \text{for} \quad x \in \bar{D}, \quad \xi \in R^n;$$

$$c_0 - \frac{1}{2}c_{0x_i}^i + \frac{1}{2}c_{0x_ix_j}^{ij} + \frac{1}{2}\left(\alpha^i a\right)_{x_i} \leq -\bar{c}_0 < 0 \ \text{for} \ x \in \bar{D};$$

$$\alpha^i a\nu_i \leq 0 \quad \text{for} \quad x \in s_1^0 \cup s_2^0$$

holds. Then, the regular solution of the boundary-value problem II is unique.

Proof. Analyzing the equality

$$-\int_D Lu \cdot u dx = -\int_D f u dx$$

and taking account of the hypotheses of the theorem and the conditions of the type obtained when proving Theorem 1.3.1 (vanishing of the derivatives u_{x_i} in the set σ_{10} and so on), we may show that, for the regular solutions of the boundary-value problem II, the following *a priori* estimate holds

$$\int_D u^2 dx \leq K_0 \int_D f^2 dx.$$

Here, the constant K_0 depends only on the coefficients of the operator L. If in equation (1.3.1) we have $f \equiv 0$, then from this estimate will follow $u \equiv 0$. The theorem is proved.

1.3.5. Existence of generalized solutions of the elliptic boundary-value problem

Suppose that the form $d^{ij,0}\xi_i\xi_j$ defined above is a negative definite form. We define the set W_0 as the set of functions $w(x)$ from the space $W_2^2(D)$ vanishing in the set $\sigma_1 \cup \sigma_2 \cup \sigma_{00} \cup s_3$ and the space H_0 as the closure of the set W_0 by the norm

$$\left|\left|w\right|\right|_{H_0} = \left[\int_D \left(d_0^{ij} w_{x_i} w_{x_j} + w^2\right) dx + \int_{\sigma_{10}} \alpha^k \nu_k a^{ij} w_{x_i} w_{x_j} d\gamma \right.$$

$$\left. - \int_{\sigma_{20}} \alpha^k \nu_k a^{ij} w_{x_i} w_{x_j} d\gamma + \int_{s_1^0 \cup s_2^0} \left|\left(c_{0x_j}^{ij} - c_0^i\right)\nu_i\right| w^2 d\gamma \right]^{\frac{1}{2}}.$$

For the functions $u(x)$ and $w(x)$ from the set W_0, we define the bilinear form

$$a^0(u, w) = \int_D \left[\alpha^k a^{ij} u_{x_i} w_{x_j x_k} + \left(\alpha^j a^{ik}\right)_{x_k} u_{x_i} w_{x_j} - \alpha^j a^i u_{x_i} w_{x_j}\right.$$

$$-c_0^{ij}u_{x_i}w_{x_j} + \left(c_{0x_j}^{ij} - c_0^i - \alpha^i a\right)uw_{x_i} + \left(c_0 - c_{0x_i}^i + c_{0x_ix_j}^{ij}\right)uw\Big]dx$$

$$+\int\limits_{s_2^0}\left(c_{0x_j}^{ij} - c_0^i\right)uw\nu_i d\gamma + \int\limits_{\sigma_{20}}\alpha^k a^{ij}u_{x_i}w_{x_k}\nu_j d\gamma.$$

Lemma 1.3.3. *For all x from the set $\bar{D}$ and ξ from R^n, let the conditions*

$$a^{ij}\xi_i\xi_j \le M_0 d_0^{ij}\xi_i\xi_j,$$

$$0 \le \left[d_0^{ij} - \frac{1}{2}\left(\alpha^k a^{ij}\right)_{x_k}\right]\xi_i\xi_j \le M_0 d_0^{ij}\xi_i\xi_j$$

hold. Then for each of the functions $u(x)$ and $w(x)$ from the set W_0 the following estimate

$$\left|a^0(u,w)\right| \le M\left|\left|u\right|\right|_{H_0}$$

holds, where the constant M depends only on the coefficients of the operator L and on the function $w(x)$.

The proof is based on the Cauchy-Schwarz-Bunyakovskii inequality and on the lemma conditions. As the proof is evident, we shall omit it.

From Lemma 1.3.3 it follows that the form $a^0(u,w)$, for $w \in W_0$ fixed, may be extended with respect to u from the set W_0 on the whole space H_0. As before, we conserve the same notation $a^0(u,w)$ for the extended form.

Definition 1.3.2. *A function $u(x)$ from the space H_0 is defined as the generalized solution of the boundary-value problem II if, for each function $w(x)$ from the set W_0, the following integral identity holds*

$$a^0(u,w) = \int\limits_D fw\,dx.$$

Theorem 1.3.6. *Let an operator A be symmetric and elliptic-parabolic in the set $\bar{D}$, an operator C_0 be elliptic-parabolic in the set σ_0^0 and suppose that the conditions*

$$d_0^{ij}\xi_i\xi_j \geq 0, \qquad a^{ij}\xi_i\xi_j \leq M_0 d_0^{ij}\xi_i\xi_j,$$

$$0 \leq \left[d_0^{ij} - \frac{1}{2}\left(\alpha^k a^{ij}\right)_{x_k}\right]\xi_i\xi_j \leq M_0 d_0^{ij}\xi_i\xi_j \text{ for } x \in \bar{D}, \ \ \xi \in R^n;$$

$$c_0 - \frac{1}{2}c^i_{0x_i} + \frac{1}{2}c^{ij}_{0x_ix_j} + \frac{1}{2}\left(\alpha^i a\right)_{x_i} \leq -\bar{c}_0 < 0 \text{ for } x \in \bar{D};$$

$$\alpha^i a\nu_i \leq 0 \quad \text{for} \quad x \in s_1^0 \cup s_2^0$$

hold. Then for each function f from the space $L_2(D)$ there exists a generalized solution of the boundary-value problem II.

The proof of this theorem, as the proof of Theorem 1.3.2, may be carried out using the Riesz theorem.

By Lemma 1.3.3 and the Riesz theorem, the form $a^0(u,w)$ generates the linear operator T defined in W_0, whose domain of values lies in H_0. This operator is defined by the equality

$$a^0(u,w) = \left(u, Tw\right)_{H_0}.$$

By the hypotheses of the theorem, for the functions w from the set W_0, we have the inequality

$$\left|a^0(u,w)\right| \geq \lambda_0\left|\left|w\right|\right|^2_{H_0}, \qquad \lambda_0 > 0.$$

This inequality yields the univalence of the operator T.

Further, as in Theorem 1.3.2, we may prove that the integral

$$\int_D fw\,dx$$

is a linear continuous functional on the space $\overline{T(W_0)}$. Again, taking into account the Riesz theorem, we obtain that there exists such an element $u(x)$ from the space H_0, which will satisfy the identity

$$\int_D fw\,dx = \Big(u, Tw\Big)_{H_0}$$

for each function $w(x)$ from the set W_0. The function $u(x)$ will be the desired generalized solution of the boundary-value problem II. The theorem is proved.

If the generalized solution $u(x)$ is a smooth function, then it will be a regular solution attaining the boundary conditions. Conversely, each regular solution of the boundary-value problem II will be a generalized solution.

1.3.6. Existence of regular solutions of the elliptic boundary-value problem

As for the boundary-value problem I, we consider the case when the operator A is an elliptic operator. First, we shall prove that in this case the generalized solution of the boundary-value problem II will have the additional smoothness and will satisfy both boundary conditions (1.3.39) and (1.3.40).

We suppose that $h(x)$ is a smooth function, positive in $D \cup \sigma_1$, equal to zero in the set $\overline{\sigma_0 \cup \sigma_2}$, and such that for each x from $\bar{D}$ the inequalities

$$\left|h_{x_i}\right| \leq K\sqrt{h}, \qquad i = 1, \ldots, n$$

hold.

Theorem 1.3.7. *Suppose that an operator A is elliptic and symmetric in $\bar{D}$ and for all x from $\bar{D}$ and ξ from R^n the following conditions hold*

$$d_0^{ij}\xi_i\xi_j \geq d_0|\xi|^2, \qquad d_0 > 0,$$

$$c_0 - \frac{1}{2}c^i_{0x_i} + \frac{1}{2}c^{ij}_{0x_ix_j} + \frac{1}{2}\Big(\alpha^i a\Big)_{x_i} \leq -\bar{c}_0 < 0.$$

Then, for the function $f(x)$ from the space $L_2(D)$, the generalized solution $u(x)$ of the boundary-value problem II will belong to the space $\overset{0}{W}{}_2^1(D)$.

Moreover, everywhere in σ_1, this solution will satisfy boundary condition (1.3.40) and the inclusion

$$\sqrt{h}\Big(l_0 u\Big)_{x_i} \in L_2(D) \quad \text{for} \quad i = 1, \ldots, n$$

will take place. If, additionally, the operator C is represented in the form

$$Cu = l_1 l_0 u + l_2 u,$$

where l_1 and l_2 are certain first order operators, then, for the generalized solution, the following inclusions will hold

$$h\Big(l_0 u\Big)_{x_i x_j} \in L_2(D); \quad \text{for} \quad i, j = 1, \ldots, n.$$

Proof. Evidently, the generalized solution $u(x)$ belongs to the space $\overset{0}{W}{}_2^1(D)$. Further, for the generalized solution, there exists a sequence $\{u^m(x)\}_{m=1}^{\infty}$ of smooth (even, infinitely differentiable) solutions such that

$$u^m \to u, \quad \text{for} \quad m \to \infty$$

in the norm of the space H_0. Moreover, using the diagonal process and considering the subsequence (where we leave the previous notation), we may establish that, for this subsequence

$$u^m_{x_i} \to u_{x_i}, \quad \text{for} \quad m \to \infty, \quad i = 1, \ldots, n$$

almost everywhere in the set $\sigma_1 \cup \sigma_2$ (in the sense of the $(n-1)$-Lebesgue measure). For the functions u^m, we denote $f_m = Lu^m$. The convergencies established above yield

$$\lim_{m \to \infty} a^0(u^m, w) = a^0(u, w) = \int\limits_D f w dx$$

for each function w from the set W_0. On the other hand, we have the equality

$$a^0(u^m, w) = \int\limits_D f_m w dx - \int\limits_{\sigma_1} \alpha^k \nu_k a^{ij} u^m_{x_i} w_{x_j} d\gamma.$$

The finite infinitely differentiable functions, evidently, belong to the set W_0. For such functions, we shall have

$$\lim_{m\to\infty} \int\limits_D (f_m - f) w dx = 0.$$

As the set of finite infinitely differentiable functions in D is dense in the space $L_2(D)$, from the last equality it follows that the sequence $\{f_m\}_{m=1}^{\infty}$ is a sequence weakly converging in the space $L_2(D)$ to the function f. Taking this into account, we obtain

$$\lim_{m\to\infty} \int\limits_{\sigma_1} \alpha^k \nu_k a^{ij} u^m_{x_i} w_{x_j} d\gamma = 0$$

for each function w from the set W_0. Hence, taking into account that $u^m_{x_i}$ converges to u_{x_i} almost everywhere, we have

$$\int\limits_{\sigma_1} \alpha^k \nu_k a^{ij} u_{x_i} w_{x_j} d\gamma = 0.$$

We set in this equality $w = u^m$ and again pass to the limit. Taking into account the convergence of derivatives almost everywhere, we obtain

$$\int\limits_{\sigma_1} \alpha^k \nu_k a^{ij} u_{x_i} u_{x_j} d\gamma = 0.$$

Positive definiteness of the quadratic form $a^{ij}\xi_i\xi_j$ gives the equalities $u_{x_i} = 0$ almost everywhere in the set σ_1 for all $i = 1, \dots, n$. Hence we obtain the desired equality $l_0 u = 0$ almost everywhere in the set σ_1. So, the first part of the theorem is proved.

Note that for the functions $u^m(x)$ the following equalities hold:

$$Lu^m = Al_0 u^m + Cu^m = f_m.$$

We now multiply the second equality on the function hl_0u^m, differentiate the result over the domain D, and integrate by parts. As a result, we obtain the equality

$$\int\limits_D ha^{ij}\left(l_0u^m\right)_{x_i}\left(l_0u^m\right)_{x_j}dx = \int\limits_D \Big[h_{x_j}a^{ij}\left(l_0u^m\right)_{x_i}l_0u^m + ha^{ij}_{x_j}\left(l_0u^m\right)_{x_i}l_0u^m$$

$$-ha^i\left(l_0u^m\right)_{x_i}l_0u^m - ha\left(l_0u^m\right)^2 + hc^{ij}u^m_{x_i}\left(l_0u^m\right)_{x_j}$$

$$+\left(hc^{ij}\right)_{x_j}u^m_{x_i}\cdot l_0u^m - hc^iu^m_{x_i}l_0u^m - hcu^ml_0u^m - hf_ml_0u^m\Big]dx$$

$$-\int\limits_{\sigma_1} ha^{ij}\left(l_0u^m\right)_{x_i}l_0u^m\nu_jd\gamma - \int\limits_{\sigma_1} hc^{ij}u^m_{x_i}l_0u^m\nu_jd\gamma. \tag{1.3.41}$$

Uniform boundedness of a family of the functions $\{u^m\}_{m=1}^{\infty}$ and $\{f_m\}_{m=1}^{\infty}$ in the spaces H_0 and $L_2(D)$ respectively, the conditions imposed on the function $h(x)$ and the Young inequality give that the integral over the domain D in the right side of the last equality is estimated from above by the value

$$\rho\int\limits_D ha^{ij}\left(l_0u^m\right)_{x_i}\left(l_0u^m\right)_{x_j}dx + K(\rho).$$

Here ρ is an arbitrary positive number; the value $K(\rho)$ is determined by ρ, the coefficients of the operator L, the domain D, and the function f. The integral over the set σ_1

$$\int\limits_{\sigma_1} ha^{ij}\left(l_0u^m\right)_{x_i}l_0u^m\nu_jd\gamma$$

may be transformed to the form without second derivatives of the functions $u^m(x)$. Really, the set σ_1 may be represented as a finite sum of smooth components. In each of these components, we may pass to the $(n-1)$-dimensional integrals and then to integrate by parts. The boundary integrals which arise will be either cancelled or will vanish (because of vanishing of the function $h(x)$). After returning to the variables $x_1,\dots,x_n$, we obtain that each integral mentioned above may be transformed to the form which does not contain second derivatives of the functions $u^m(x)$.

As in the set σ_1 we have that $u^m_{x_i} \to 0$ almost everywhere, we obtain that both boundary integrals in the right side of equality (1.3.41) tend to zero for $m \to \infty$.

Choosing the constant ρ sufficiently small, we obtain the estimate

$$\sum_{i=1}^{n} \int_D h\left[\left(l_0 u^m\right)_{x_i}\right]^2 dx \leq M_0 \tag{1.3.42}$$

with the constant M_0 dependent only on the coefficients of the operator L, the domain D, and the function f. This estimate will be valid if we pass to the limit for $m \to \infty$. In other words, for the function $u(x)$, the estimate (1.3.42) holds; thus, we have the inclusion

$$\sqrt{h}\left(l_0 u^m\right)_{x_i} \in L_2(D).$$

Finally, to establish the third part of the theorem, we use the following identity

$$A\left(hl_0 u^m\right) = hf_m - hCu^m + a^{ij}h_{x_j}\left(l_0 u^m\right)_{x_j} + a^i h_{x_i} l_0 u^m + Ahl_0 u^m.$$

Taking account of the special form of the operator C, the uniform estimate of the set of functions $\{u^m\}_{m=1}^{\infty}$ in the space H_0 and estimate (1.3.42), we see that the right side of this equality is bounded in $L_2(D)$ uniformly with respect to m. Applying now the considerations similar to those when we have obtained the second basic inequality for elliptic operators (Ladyzhenskaya and Ural'tseva, 1973) and passing to the limit, we obtain the required inclusions. The theorem is proved.

Now, let $h_1(x)$ be a smooth function nonnegative in $\bar{D}$, equal to zero in σ_2 and such that for each x from $\bar{D}$, we have the inequalities

$$\left|h_{1x_i}\right| \leq K\sqrt{h_1}, \qquad i = 1, \ldots, n.$$

Introduce the following notation

$$d^{ij,1} = \left(\alpha - \frac{1}{2}\alpha^k_{x_k}\right) a^{ij} + b^{ij}.$$

Theorem 1.3.8. *Let A be an elliptic symmetric operator in $\bar{D}$; suppose that the conditions*

$$d_0^{ij}\xi_i\xi_j \geq d_0|\xi|^2, \quad d_1^{ij}\xi_i\xi_j \geq d_1|\xi|^2, \quad d_0 > 0, \quad d_1 > 0,$$

$$c_0 - \frac{1}{2}c^i_{0x_i} + \frac{1}{2}c^{ij}_{0x_ix_j} + \frac{1}{2}\left(\alpha^i a\right)_{x_i} \leq 0$$

hold for all x from $\bar{D}$ and ξ from R^n; and suppose that

$$\alpha^k \nu_k \leq 0 \quad \text{for all} \quad x \in \Gamma.$$

Then, for each function $f(x)$ from the space $L_2(D)$, there exists a solution $u(x)$ of the boundary-value problem II such that

$$u(x) \in W_2^2(D) \bigcap \overset{0}{W}{}_2^1(D), \qquad h_1 \cdot l_0 u \in W_2^2(D).$$

Proof. For $\varepsilon > 0$, we consider the problem: *find a solution of the equation*

$$L_\varepsilon u = -\varepsilon A^* A u + L u = f, \tag{1.3.43}$$

satisfying the conditions

$$u\Big|_\Gamma = 0, \qquad Au\Big|_\Gamma = 0. \tag{1.3.44}$$

Here A^* is the formally conjugate operator to A.

Boundary-value problem (1.3.43), (1.3.44) has the solution $u^\varepsilon(x)$ from the space $W_2^4(D)$ (Agmon *et al.*, 1962). We show now that the set of functions $\{u^\varepsilon(x)\}$ satisfies the a *priori* estimates which allow us to pass to the limit and to obtain the required solution of the boundary-value problem II.

The equality

$$-\int_D L_\varepsilon u^\varepsilon \cdot u^\varepsilon dx = -\int_D f u^\varepsilon dx$$

yields the first *a priori* estimate

$$\varepsilon \int_D \left(Au^\varepsilon\right)^2 dx + \left\|u^\varepsilon\right\|^2_{\overset{\circ}{W}{}_2^1(D)} \leq M_1, \tag{1.3.45}$$

where the constant M_1 depends only on the coefficients of the operator L and on the function f.

The following equality

$$\int_D L_\varepsilon u^\varepsilon \cdot Au^\varepsilon dx = \int_D f \cdot Au^\varepsilon dx$$

we integrate by parts. Taking account of the hypotheses of the theorem and estimate (1.3.45), we obtain the estimate

$$\varepsilon \sum_{i=1}^n \int_D \left[\frac{\partial}{\partial x_i}\left(Au^\varepsilon\right)\right]^2 dx + \left\|Au^\varepsilon\right\|^2_{L_2(D)} \leq M_2. \tag{1.3.46}$$

Here the constant M_2 again depends only on the coefficients of the operator L and the function f.

The second basic inequality for the elliptic operators together with estimates (1.3.45) and (1.3.46) yield the estimate

$$\left\|u^\varepsilon\right\|^2_{W_2^2(D)} \leq M_3 \tag{1.3.47}$$

with the constant M_3 dependent on the coefficients of the operator L, the domain D, and the function f.

Now, we consider the equality

$$\int_D L_\varepsilon u^\varepsilon h_1^2 l_0 Au^\varepsilon dx = \int_D f \cdot h_1^2 l_0 Au^\varepsilon dx.$$

Integrating by parts we transform this equality as follows:

$$\int_D h_1^2\left(l_0 Au^\varepsilon\right)^2 dx - \frac{\varepsilon}{2}\int_\Gamma h_1^2 a^{ij}\left(Au^\varepsilon\right)_{x_i}\left(Au^\varepsilon\right)_{x_j} \alpha^k \nu_k d\gamma$$

$$+\varepsilon \int_\Gamma h_1^2 a^{ij}\left(Au^\varepsilon\right)_{x_i} \alpha^k \left(Au^\varepsilon\right)_{x_k} \nu_j d\gamma = I, \tag{1.3.48}$$

Here the value I standing in the right side is the finite sum of the integrals over the domain D. Taking account of estimates (1.3.45)-(1.3.47) and the conditions imposed on the function h_1, we see that each of these integrals is bounded by a certain function dependent on the constants $M_1 - M_3$ and the function f. As the equality

$$\left(Au^{\varepsilon}\right)_{x_k}\nu_j = \left(Au^{\varepsilon}\right)_{x_j}\nu_k$$

holds almost everywhere in Γ, then the boundary integrals, standing in the left side of equality (1.3.48), vanish. Therefore, equality (1.3.48) yields the estimate

$$\int\limits_D h_1^2\left(l_0 Au^{\varepsilon}\right)^2 dx \leq M_4, \tag{1.3.49}$$

where the constant M_4 (similar to the constants $M_1 - M_3$) depends only on the coefficients of the operator L, the domain D, and the function f.

For the functions u^{ε}, the following equality

$$A\left(h_1 l_0 u^{\varepsilon}\right) = h_1 l_0 Au^{\varepsilon} + \Phi u^{\varepsilon}$$

holds, where Φ is a certain differential operator of second order. The right side of this equality belongs to the space $L_2(D)$; therefore, the left side of this equality is also a function which belongs to $L_2(D)$. The function $h_1 l_0 u^{\varepsilon}$ vanishes in the boundary of D. Therefore, from the second basic inequality for elliptic operators, it follows that the function $h_1 l_0 u^{\varepsilon}$ belongs to the space $W_2^2(D)$. Moreover, taking account of estimate (1.3.47), we see that, for this function, the uniform estimate will hold

$$\left|\left|h_1 l_0 u^{\varepsilon}\right|\right|^2_{W_2^2(D)} \leq M_5. \tag{1.3.50}$$

From estimates (1.3.45), (1.3.47), and (1.3.50) it follows that from the set of functions $\{u^{\varepsilon}(x)\}$ we may select a subsequence $\{u^{\varepsilon_m}(x)\}_{m=1}^{\infty}$ such that, for a certain function $u(x)$, we shall have

$$u^{\varepsilon_m} \to u \quad \text{weakly in} \quad W_2^2(D),$$

$$h_1 l_0 u^{\varepsilon_m} \to h_1 l_0 u \quad \text{weakly in} \quad W_2^2(D), \tag{1.3.51}$$

for $m \to \infty$. Moreover, for the limit functions, the estimates (1.3.45), (1.3.47), and (1.3.50) will hold.

In equation (1.3.43), set $\varepsilon = \varepsilon_m$ and multiply this equation by a function $w(x)$ from the space $W_2^2(D) \cap \overset{0}{W}{}_2^1(D)$. After integrating we obtain the equality

$$-\varepsilon_m \int_D Au^{\varepsilon_m} \cdot Aw dx + a^0\left(u^{\varepsilon_m}, w\right) = \int_D fw dx.$$

Taking into account estimate (1.3.45), we see that the first term of the left side of this equality tends to zero. Taking account of convergence (1.3.51), we obtain the equality for the function $u(x)$

$$a^0(u, w) = \int_D fw dx.$$

As the function $w(x)$ is arbitrary, the function $u(x)$ will be a generalized solution of the boundary-value problem II. On the other hand, for the function $u(x)$, we shall have the inclusions:

$$u \in W_2^2(D) \bigcap \overset{0}{W}{}_2^1(D), \qquad h_1 l_0 u \in W_2^2(D).$$

Therefore, the function $u(x)$ will be the desired solution of the problem. The theorem is proved.

The set σ_1 is empty if the conditions of Theorem 1.3.8 hold. The boundary-value problem II, in this case, becomes the Dirichlet problem; otherwise, when the set σ_1 is the whole boundary Γ, the boundary-value problem II coincides with the boundary-value problem I, and the existence of regular solutions follows from Theorem 1.3.3.

In conclusion, as in Section 3.3, we shall state some results on existence of regular solutions for equations with the selected variable.

In the cylindrical domain $Q = D \times (0, T)$ of a finite height T, we consider the equation

$$L_\varepsilon u \equiv \left(K(t)\frac{\partial}{\partial t} + \alpha(t)\right)\left(u_{tt} + Au\right) + Bu = f(x,t) \tag{1.3.52}$$

where the operators A and B are as follows:

$$Au = \frac{\partial}{\partial x_i}\left(a^{ij}u_{x_j}\right) - a_0(x)u, \quad Bu = \frac{\partial}{\partial x_i}\left(b^{ij}u_{x_j}\right) - b_0(x)u,$$

and the operator A is elliptic for $x \in \bar{D}$, the operator B is elliptic-parabolic for $x \in \bar{D}$. We shall assume, as usual, that the coefficients of the operator L are smooth functions for $x \in \bar{D}$, $t \in [0,T]$; the function $K(t)$ is assumed to be nonnegative:

$$K(t) \geq 0 \text{ for } t \in (0,T); \quad K(0) > 0, \quad K(T) > 0.$$

In this case, the set σ_1 is the lower base of the cylinder Q; the set σ_2 is the upper base of the cylinder Q; and the set σ_0 is the lateral area of the cylinder Q. The boundary-value problem becomes as follows: find a solution of equation (1.3.52) satisfying the conditions

$$u\Big|_{\partial Q} = 0, \tag{1.3.53}$$

$$u_t(x,0) = 0. \tag{1.3.54}$$

We shall show now that boundary-value problem (1.3.52)-(1.3.54) has an almost regular solution.

Theorem 1.3.9. *Suppose the conditions imposed on the operators A and B hold, the operator A is symmetric in $\bar{D}$, and, for $x \in \bar{D}$, $t \in [0,T]$, the following conditions hold*

$$a_0(x) \geq 0, \qquad b_0(x) \geq 0,$$

$$\alpha(t) - \frac{p}{2}K_t(t) \geq \bar{\alpha}_0 > 0, \qquad p = 1,3;$$

$$a_0\left(\alpha - \frac{1}{2}K_t\right) - \frac{1}{2}\alpha_{tt} + \frac{1}{2}K_{ttt}(t) \geq \bar{\alpha}_1 > 0.$$

Then, for each function $f(x,t)$ such that $f(x,t)$, $f_t(x,t)$ belong to the space $L_2(Q)$, there exists the solution $u(x,t)$ of the boundary-value problem (1.3.52)-(1.3.54) such that the following inclusions

$$u \in W_2^2(Q)\bigcap \overset{0}{W}{}_2^1(Q), \qquad \sqrt{(T-t)K(t)}u_{ttt} \in L_2(Q),$$

$$\sqrt{(T-t)K(t)}u_{x_ix_jt} \in L_2(Q), \quad i,j = 1,\dots,n$$

and condition (1.3.54) will take place.

Proof. We shall use the regularization method. For $\varepsilon > 0$, we consider the set of equations

$$L_\varepsilon u \equiv \left(K_\varepsilon(t)\frac{\partial}{\partial t} + \alpha(t)\right)\left(u_{tt} + Au\right) + Bu = f(x,t), \qquad (1.3.52_\varepsilon)$$

where $K_\varepsilon = K + \varepsilon$, and the set of the corresponding boundary-value problems: *find the solution of equation ($1.3.52_\varepsilon$) satisfying conditions (1.3.53) and (1.3.54).* In its turn, the establishment of the existence of a good solution of boundary-value problem ($1.3.52_\varepsilon$), (1.3.53), (1.3.54) will be based on the Galyerkin method and the method of extension by parameter. For both these methods and for the passing to the limit by the regularization parameter we need the *a priori* estimates; therefore, first, we show that for solutions of problem ($1.3.52_\varepsilon$), (1.3.53), (1.3.54), the uniform norm estimates hold, which correspond to the theorem inclusions.

So, we suppose that $u(x,t)$ solves problem ($1.3.52_\varepsilon$), (1.3.53), (1.3.54).

The estimate

$$\|u\|_{\overset{0}{W}{}_2^1(Q)} \leq N_1 \qquad (1.3.55)$$

is evident for the function $u(x,t)$, where the constant N_1 depends only the coefficients of the operator L and the norm of the function $f(x,t)$ in the space $L_2(Q)$.

To obtain the second estimate, we consider the equality

$$\int_Q L_\varepsilon u \cdot A_0 u dxdt = \int_Q f \cdot A_0 u dxdt,$$

where A_0 is the operator $A + a_0$. After integrating by parts, we obtain the equality

$$\int_Q \left(\alpha - \frac{1}{2}K_t\right)\left(A_0 u\right)^2 dxdt + \int_Q \left(\alpha - \frac{3}{2}K_t\right) a^{ij} u_{x_i t} u_{x_j t} dxdt$$

$$+\frac{1}{2}\int_D K_\varepsilon a^{ij} u_{x_i t}(x,T) u_{x_j t}(x,T) dx + \int_Q A_0 u \cdot Bu dxdt$$

$$= \int_Q f \cdot A_0 u dxdt + \frac{1}{2}\int_Q \left(\alpha_{tt} - K_{ttt}\right) a^{ij} u_{x_i} u_{x_j} dxdt$$

$$- \int_Q \alpha a_0 u \cdot A_0 u dxdt - \int_Q K_\varepsilon a_0 u_t \cdot A_0 u dxdt. \tag{1.3.56}$$

The first term of the left side of equality (1.3.56) is estimated from below taking account of estimate (1.3.55)

$$\int_Q \left(\alpha - \frac{1}{2}K_t\right)\left(A_0 u\right)^2 dxdt \geq k_0 \sum_{i,j=1}^{n} \int_Q u_{x_i x_j}^2 dxdt - k_1.$$

Here the constants k_0 and k_1 are positive and depend only on the coefficients of the operator L and the domain D. Further, using the ideas of the proof of the basic inequality for elliptic operators, we may establish the following inequality

$$\int_Q A_0 u \cdot Bu dxdt \geq -k_2.$$

The constant k_2 here is positive and depends only on the coefficients of the operator L and the domain D (here we use also estimate (1.3.55)). Finally, the right side of equality (1.3.56) is estimated from above by the value

$$\rho \sum_{i,j=1}^{n} \int_{Q} u_{x_i x_j}^2 \, dx dt + N_2(\rho).$$

Here ρ is an arbitrary positive number, the constant N_2 depends on ρ, the coefficients of the operator L, and the domain D.

From all these inequalities and the hypotheses of the theorem it follows that

$$\sum_{i,j=1}^{n} \int_{Q} u_{x_i x_j}^2 \, dx dt + \sum_{i=1}^{n} \int_{Q} u_{x_i t}^2 \, dx dt + \sum_{i=1}^{n} \int_{D} K_\varepsilon(T) u_{x_i t}^2(x,T) \, dx \leq N_3, \tag{1.3.57}$$

where the constant N_3 depends only on the coefficients of the operator L, the domain D, and the norm of the function $f(x,t)$ in the space $L_2(Q)$.

We suppose that $\varphi(t)$ is a smooth function in the interval $[0,T]$ and such that $\varphi(T) = \varphi'(T) = 0$; $\varphi(0) = 1$. Consider now the equality

$$\int_0^T L_\varepsilon u \cdot \varphi(t) \, dt = \int_0^T f \varphi(t) \, dt.$$

Representing the integrand from the left side of this equality in the form

$$L_\varepsilon u \cdot \varphi = \frac{\partial}{\partial t}\Big[K_\varepsilon \varphi u_{tt} - \Big(K_\varepsilon \varphi\Big)_t u_t + K_\varepsilon \varphi A u + \alpha \varphi u_t\Big]$$

$$+ \Big(K_\varepsilon \varphi\Big)_{tt} u_t - \Big(\alpha \varphi\Big)_t u_t + \alpha \varphi A u + \varphi B u$$

and integrating, we obtain the equality

$$K_\varepsilon u_{tt}(x,0) = \int_0^T \Big\{\Big[\Big(K_\varepsilon \varphi\Big)_{tt} - \Big(\alpha \varphi\Big)_t\Big] u_t + \alpha \varphi A u + \varphi B u\Big\} dt - \int_0^T \varphi f \, dt.$$

We now multiply this equality from the right and from the left by $u_{tt}(x,0)$ and integrate over the domain D. Taking account of the Hölder inequality and estimates (1.3.55), (1.3.57), we obtain the estimate

$$\int\limits_D u_{tt}^2(x,0)dx \leq N_4, \tag{1.3.58}$$

where the constant N_4 depends only on the coefficients of the operator L, the domain D, the number T, and the norm of the function $f(x,t)$ in the space $L_2(Q)$.

Taking into account estimates (1.3.55), (1.3.57), (1.3.58), and the equality

$$\int\limits_Q L_\varepsilon u \cdot u_{tt} dxdt = \int\limits_Q f u_{tt} dxdt,$$

we establish the estimate

$$\int\limits_Q u_{tt}^2 dxdt \leq N_5. \tag{1.3.59}$$

In this estimate, the constant N_5, like the constant N_4, depends only on the coefficients of the operator L, the domain D, the number T, and the $L_2(Q)$-norm of the function $f(x,t)$.

To obtain the last estimate, we consider the equality

$$\int\limits_Q (T-t)L_\varepsilon u \cdot u_{ttt} dxdt = \int\limits_Q (T-t) f u_{ttt} dxdt.$$

For a fixed ε, estimates (1.3.55), (1.3.57)-(1.3.59) yield the estimate

$$\int\limits_Q (T-t)u_{ttt}^2 dxdt \leq N_6, \tag{1.3.60}$$

where the constant N_6 depends only on the coefficients of the operator L, the domain D, the number T, the $L_2(Q)$-norm of the function $f(x,t)$, and on the number ε. To obtain the estimate uniform with respect to ε, it is necessary to integrate by parts in the right side of the last equality. Again, taking account of estimates (1.3.55), (1.3.57)-(1.3.59), we obtain the estimate

$$\int_Q (T-t)K_\varepsilon(t)u_{ttt}^2\,dxdt \le N_7. \tag{1.3.61}$$

Here the constant N_7 depends on the coefficients of the operator L, the domain D, the number T, and the $L_2(Q)$-norm of the functions $f(x,t)$ and $f_t(x,t)$.

All the necessary estimates are obtained. Hence it follows that the solutions of problems (1.3.52$_\varepsilon$), (1.3.53), (1.3.54) and (1.3.52)-(1.3.54) exist. To establish the existence of solutions to problem (1.3.52$_\varepsilon$), (1.3.53), (1.3.54), we consider the one-parameter set of the operators

$$L_{\varepsilon,\tau}u \equiv \left(K_\varepsilon\frac{\partial}{\partial t}+\alpha\right)\left(u_{tt}+Au\right)+\tau Bu.$$

Here τ is an arbitrary number from the interval $[0,1]$. The setting of the corresponding boundary-value problems is as follows: *find the solution of the equation*

$$L_{\varepsilon,\tau}u = f(x,t), \tag{1.3.52$_{\varepsilon,\tau}$}$$

*satisfying conditions (1.3.53), (1.3.54).*When $\tau=0$, this problem has a regular solution. More precisely, the solution which satisfies the inclusions

$$u\in W_2^2(Q)\bigcap \overset{0}{W}{}_2^1(Q),\qquad \sqrt{T-t}\,u_{ttt}\in L_2(Q),$$

$$\sqrt{T-t}\,u_{x_ix_jt}\in L_2(Q),\quad i,j=1,\dots,n.$$

This may be shown if we use the Galyerkin method and estimates (1.3.55), (1.3.57)-(1.3.60). As estimates (1.3.55), (1.3.57)-(1.3.60) are established for the functions $f(x,t)$ from the space $L_2(Q)$, then we may apply to the set of boundary-value problems (1.3.52$_{\varepsilon,\tau}$), (1.3.53), (1.3.54), for fixed ε, the method of extension by the parameter. Therefore, boundary-value problem (1.3.52$_{\varepsilon,\tau}$), (1.3.53), (1.3.54) has the solution which satisfies estimates (1.3.55), (1.3.57)-(1.3.60) for all τ from the interval $[0,1]$. In particular,

$\tau = 1$ also enters this interval. Finally, estimates (1.3.55), (1.3.57)-(1.3.59), (1.3.61), uniform with respect to ε, allow us to pass to the limit for $\varepsilon \to 0$ and to obtain the required solution of boundary-value problem (1.3.52)-(1.3.54). The theorem is proved.

Equation (1.3.52), like equation (1.3.23) considered in point 3.3, is the model equation. In particular, Theorem 1.3.9 holds for equations of the type (1.3.52) with an arbitrary elliptic symmetric operator A of the form

$$Au = a^{ij}(x)u_{x_i x_j} + a^i(x)u_{x_i} + a_0(x)u,$$

and the operator B of the form

$$Bu = B_0 u + B_1 u.$$

Here B_0 is an elliptic-parabolic operator of the general form and B_1 is a differential operator of second order of the general type satisfying a certain proximity condition. The corresponding conditions for the coefficients of the operators A and B may be easily written. The existence of the required solution may be established using the method of continuation by parameter considering the set of the operators

$$L_\tau u \equiv \left(K(t)\frac{\partial}{\partial t} + \alpha(t)\right)\left(u_{tt} + \tau Au + (1-\tau)\Delta_x u\right) + \tau Bu$$

(here Δ_x is the Laplace operator with respect to the variables $x_1, \ldots, x_n$) and the set of the corresponding boundary-value problems. Further, the functions K and α may depend on the variable x; and the coefficients of the operators A and B may depend on the variable t. Finally, if the function K vanishes either for $t = 0$, or for $t = T$, or for $t = 0$ and $t = T$, then at most one of the sets σ_1 and σ_2 is empty. The investigation of the corresponding boundary-value problem is simpler than those of boundary-value problem (1.3.52)-(1.3.54).

We consider now one more class of equations of composite type with the selected variable.

Namely, in the cylinder Q, we consider the equation

$$Lu \equiv u_{tt} - Au_t + Bu = f(x,t), \tag{1.3.62}$$

where the operator A is as follows:

$$Au = \frac{\partial}{\partial x_i}\Big(a^{ij}(x)u_{x_j}\Big) - a_0(x)u$$

and is elliptic in $\bar{D}$. The operator B is the general operator of second order

$$Bu = b^{ij}u_{x_ix_j} + b^i(x)u_{x_i} + b_0(x)u.$$

According as the setting of the boundary-value problem II, we consider the following boundary-value problem: *find the solution of equation (1.3.62) satisfying the condition*

$$u\Big|_{\partial Q} = 0. \tag{1.3.63}$$

Theorem 1.3.10. *Suppose that for a certain positive constant λ_0, for all x from $\bar{D}$ and ξ from R^n, the conditions*

$$\lambda_0 a_0 - 2b_0 - b^{ij}_{x_ix_j} + b^i_{x_i} - \lambda_0^2 \geq \bar{a}_0 > 0,$$

$$\Big(\lambda_0 a^{ij} + 2b^{ij}\Big)\xi_i\xi_j \geq \bar{a}_1|\xi|^2, \quad \bar{a}_1 > 0$$

hold. If a function $f(x,t)$ belongs to the space $L_2(Q)$, then problem (1.3.62), (1.3.63) will have a solution $u(x,t)$ such that

$$u \in L_2\Big(0,T; W_2^2(D)\Big) \bigcap \overset{0}{W}{}_2^1(Q),$$

$$u_t \in L_2\Big(0,T; \overset{0}{W}{}_2^1(D)\Big) \bigcap L_2\Big(\rho,T; W_2^2(D)\Big), \qquad \sqrt{t}u_{tt} \in L_2(Q),$$

where ρ is an arbitrary positive number less than T.

Proof. We again shall use the regularization method. For $\varepsilon > 0$, we consider the set of equations

$$L_\varepsilon u \equiv -\varepsilon A_0 u_{tt} + Lu = f(x,t), \tag{1.3.62$_\varepsilon$}$$

where A_0 is the higher part of the operator A, and the set of the boundary-value problems: *find the solutions of the equation (1.3.62$_\varepsilon$) satisfying condition (1.3.63).* The existence of regular solutions of boundary-value problem (1.3.62$_\varepsilon$), (1.3.63) is known (see, for example, Dubinskii, 1973). We shall establish the *a priori* estimate which we need for passing to the limit.

We suppose that $u(x,t)$ is a regular solution of the problem (1.3.62$_\varepsilon$), (1.3.63) (the index ε we shall not write temporarily). The equality

$$-\int_Q L_\varepsilon u \cdot u \cdot \exp\left(\lambda_0 t\right) dxdt = -\int_Q fu \cdot \exp\left(\lambda_0 t\right) dxdt$$

yields the first *a priori* estimate

$$\varepsilon \left\| u_t \right\|^2_{L_2(0,T;W_2^1(D))} + \left\| u \right\|^2_{W_2^1(Q)} \le M_1 \tag{1.3.64}$$

with the constant M_1 dependent only on the coefficients of the operator L, the number T, and the function $f(x,t)$.

Now, we consider the equality

$$\int_Q L_\varepsilon u \cdot A_0 u \cdot \exp\left(\lambda_0 t\right) dxdt = \int_Q f A_0 u \cdot \exp\left(\lambda_0 t\right) dxdt.$$

Integrating by parts, we reduce this equality to the form

$$\varepsilon \int_Q \left(A_0 u_t\right)^2 \exp\left(\lambda_0 t\right) dxdt$$

$$+ \int_Q \left(\frac{\lambda_0}{2} A_0 u + B_0 u - \frac{\varepsilon \lambda_0^2}{2} A_0 u\right) \cdot A_0 u \exp\left(\lambda_0 t\right) dxdt$$

$$+\int\limits_Q a^{ij}u_{x_i t}u_{x_j t}\exp\Big(\lambda_0 t\Big)dxdt = \int\limits_Q f\cdot A_0u\cdot\exp\Big(\lambda_0 t\Big)dxdt$$

$$+\int\limits_Q a_0u_t\cdot A_0u\cdot\exp\Big(\lambda_0 t\Big)dxdt - \int\limits_Q b^iu_{x_i}A_0u\cdot\exp\Big(\lambda_0 t\Big)dxdt$$

$$-\int\limits_Q b_0u\cdot A_0u\cdot\exp\Big(\lambda_0 t\Big)dxdt + \frac{\lambda^2}{2}\int\limits_Q a^{ij}u_{x_i}u_{x_j}\exp\Big(\lambda_0 t\Big)dxdt. \qquad (1.3.65)$$

Here B_0u is the higher part of the operator B

$$B_0u = b^{ij}u_{x_i}u_{x_j}.$$

Note that there exists a positive number ε_0 such that, for $\varepsilon < \varepsilon_0$, the operator

$$\frac{\lambda_0}{2}A_0u + B_0u - \frac{\varepsilon\lambda_0^2}{2}A_0u$$

will be elliptic. In this case, the second term from the left side of equality (1.3.65) may be estimated from below as follows:

$$\int\limits_Q\Big(\frac{\lambda_0}{2}A_0u + B_0u - \frac{\varepsilon\lambda_0^2}{2}A_0u\Big)\cdot A_0u\exp\Big(\lambda_0 t\Big)dxdt$$

$$\geq k_0\sum_{i,j=1}^n\int\limits_Q u^2_{x_ix_j}\exp\Big(\lambda_0 t\Big)dxdt - k_1\Big|\Big|u\Big|\Big|_{L_2(0,T;W_2^1(D))}.$$

Here the constants k_0 and k_1 are positive and depend on the coefficients of the operator L and the domain D. Taking this equality into account, estimating the right side of equality (1.3.65) by the Young inequality and using estimate (1.3.64), we obtain the estimate

$$\varepsilon\int\limits_Q\Big(A_0u\Big)^2dxdt + \sum_{i,j=1}^n\int\limits_Q u^2_{x_ix_j}dxdt + \sum_{i=1}^n\int\limits_Q u^2_{x_it}dxdt \leq M_2. \qquad (1.3.66)$$

The constant M_2 in this inequality depends on the coefficients of the operator L, the domain D, the number T and the function $f(x,t)$.

The next estimate follows from the analysis of the equality

$$\int_Q L_\varepsilon u \cdot t u_{tt} dxdt = \int_Q f \cdot t u_{tt} dxdt.$$

Namely, taking into account this equality and estimates (1.3.64), (1.3.66), we obtain the estimate

$$\int_Q t u_{tt}^2 dxdt \leq M_3, \tag{1.3.67}$$

where the constant M_3, like the constant M_2, depends only on the coefficients of the operator L, the domain D, the number T, and the function $f(x,t)$.

Finally, the equality

$$-\int_Q L_\varepsilon u \cdot t A_0 u_t dxdt = -\int_Q f \cdot t A_0 u_t dxdt$$

and estimates (1.3.64), (1.3.66), and (1.3.67) yield the estimate

$$\int_Q t\Big(A_0 u_t\Big)^2 dxdt \leq M_4 \tag{1.3.68}$$

with the constant M_4 dependent on the same values as the constants M_1 and M_3.

Estimates (1.3.64), (1.3.66)-(1.3.68) are sufficient for passing to the limit and for establishing the generalized solution of problem (1.3.62), (1.3.63) (passing to the limit is performed in the standard way). Moreover, for the limit functions, all the estimates above taken for $\varepsilon = 0$ will hold true; therefore, all the desired inclusions will hold. The theorem is proved.

In the next point we show that, in certain special cases, the condition of ellipticity may be weakened.

1.3.7. Correlation of the hyperbolic and elliptic boundary-value problems

The boundary-value problems I and II were defined above as hyperbolic and elliptic boundary-value problems respectively. Really, from the setting of these problems it follows that the leading role in the setting of the boundary-value problem I plays the operator l_0. This operator is the first order hyperbolic operator. Moreover, when we have proved the existence of solutions, we have used the properties of solutions of the corresponding boundary-value problem for the operators of the first order. In the setting of the boundary-value problem II, the role of the hyperbolic operator l_0 is essentially smaller. Further, the boundary-value problem I, in the case when equation (1.3.1) is a hyperbolic equation of the second order or is either a pseudohyperbolic or a pseudoparabolic equation of the third order, coincides with the initial boundary-value problem. The boundary-value problem II, for example, for hyperbolic equations of the second order, gives the ill-posed problem. On the other hand, the boundary-value problem II is the most natural for the classical equations of composite type

$$\frac{\partial}{\partial x_1}\Delta u + c(x)u = f(x).$$

Both the setting of the boundary-value problem II and the methods of its investigation for this equation follow first of all from the properties of the Laplace operator. Finally, the most essential argument is that the boundary-value problems I and II in certain cases have such duality as have the Dirichlet problem for hyperbolic equations and the Cauchy problem for an elliptic equation.

We shall establish the last statement.

Suppose that the domain D is the Cartesian product of the domains D' and D''

$$D = D' \times D'', \quad D' \subset R^m, \quad D'' \subset R^{n-m}, \quad 1 \leq m < n.$$

Suppose that $x' = (x_1, \dots, x_m)$ is an arbitrary point from the domain D' and $x'' = (x_{m+1}, \dots, x_n)$ is an arbitrary point from the domain D''.

In the cylinder $Q = D \times (0, T)$ we consider the equation

$$L_\kappa u \equiv u_{tt} - Au_t - Bu + \kappa Cu = f(x,t), \tag{1.3.69}$$

where the parameter κ attains one of the three values $\kappa = 0$, $\kappa = -1$, or $\kappa = 1$. The operators A, B, and C are defined as follows:

$$Au \equiv \sum_{i,j=1}^{m} \frac{\partial}{\partial x_i}\Big(a^{ij}(x')u_{x_j}\Big) - a_0(x')u,$$

$$Bu \equiv \sum_{i,j=1}^{m} b^{ij}(x')u_{x_i x_j} + \sum_{i=1}^{m} b^i(x')u_{x_i} + b_0(x')u,$$

$$Cu \equiv \sum_{i,j=m+1}^{n} \frac{\partial}{\partial x_i}\Big(c^{ij}(x'')u_{x_j}\Big) - c_0(x'')u.$$

Here the operator B is an arbitrary operator, and the operators A and B satisfy the conditions

$$\sum_{i,j=1}^{m} a^{ij}(x')\xi_i\xi_j \geq k_0 \cdot \Big(\xi_1^2 + \ldots + \xi_m^2\Big), \quad k_0 > 0,$$

$$\sum_{i,j=m+1}^{n} c^{ij}(x'')\xi_i\xi_j \geq k_1 \cdot \Big(\xi_{m+1}^2 + \ldots + \xi_n^2\Big), \quad k_1 > 0,$$

for each of the points x' and x'' from the sets $\bar{D}'$ and $\bar{D}''$ respectively and for each of the vectors $(\xi_1, \ldots, \xi_m)$ and $(\xi_{m+1}, \ldots, \xi_n)$ from the spaces R^m and R^{n-m} respectively.

For $\kappa = 0$, equation (1.3.69) we shall consider as the equation in the cylinder $\Omega = D' \times (0, T)$. The boundary-value problem I, for equations (1.3.69), we shall consider as the first initial boundary-value problem: *find the solution of equation (1.3.69) satisfying the conditions*

$$u(x, 0) = u_0(x), \qquad u_t(x, 0) = u_1(x) \tag{1.3.70}$$

and the condition

$$u\Big|_{\partial D \times (0,T)} = 0, \tag{1.3.71}$$

and, for $\kappa = 0$, the condition

$$u\Big|_{\partial D'\times(0,T)} = 0. \tag{1.3.71$_0$}$$

The boundary-value problem II, as earlier, we shall consider as the problem with homogeneous boundary conditions.

Theorem 1.3.11. *Suppose in equation (1.3.69) we have $\kappa = 0$. If the coefficient $a_0(x')$ is strictly positive in the set $\bar{D}'$ and the functions f, u_0, and u_1 are such that*

$$f \in L_2(\Omega), \quad u_0 \in W_2^2(D')\bigcap \overset{0}{W}{}_2^1(D'), \quad u_1 \in \overset{0}{W}{}_2^1(D'),$$

then boundary-value problem (1.3.69), (1.3.70), (1.3.71$_0$) (i.e. the boundary-value problem I) has a unique solution $u(x,t)$ such that it satisfies the inclusions

$$u \in L_\infty\Big(0, T; W_2^2(D')\bigcap \overset{0}{W}{}_2^1(D')\Big),$$

$$u_t \in L_2\Big(0, T; W_2^2(D')\Big)\bigcap L_\infty\Big(0, T; W_2^2(D')\Big), \quad u_{tt} \in L_2(\Omega).$$

The homogeneous Dirichlet problem (i.e. the boundary-value problem II), for similar conditions imposed on $f(x,t)$ and $a_0(x')$, has the solution $u(x,t)$ such that

$$u \in L_2\Big(0, T; W_2^2(D')\Big)\bigcap \overset{0}{W}{}_2^1(\Omega),$$

$$u_t \in L_2\Big(0, T; \overset{0}{W}{}_2^1(D')\Big)\bigcap L_2\Big(\rho, T; W_2^2(D')\Big), \quad \sqrt{t}u_{tt} \in L_2(\Omega),$$

where the number ρ is an arbitrary positive number less than T.

If in equation (1.3.69) we have $\kappa = -1$, the coefficients $a_0(x')$ and $c_0(x'')$ are strictly positive in the sets $\bar{D}'$ and $\bar{D}''$ respectively, and the functions f, u_0, and u_1 are such that

$$f, f_t \in L_2(Q), \quad u_0 \in W_2^2(D) \bigcap \overset{0}{W}{}_2^1(D), \quad u_1 \in W_2^2(D) \bigcap \overset{0}{W}{}_2^1(D),$$

then boundary-value problem (1.3.69), (1.3.70), (1.3.71) has a unique solution $u(x,t)$ such that

$$u \in L_\infty\Big(0,T; W_2^2(D) \bigcap \overset{0}{W}{}_2^1(D)\Big),$$

$$u_t \in L_\infty\Big(0,T; W_2^2(D) \bigcap \overset{0}{W}{}_2^1(D)\Big), \qquad u_{tt} \in L_\infty(Q).$$

The boundary-value problem II may have not a unique solution if $\kappa = -1$.

If in equation (1.3.69) we have $\kappa = 1$, the coefficients $a_0(x')$ and $c_0(x'')$ are strictly positive in the sets $\bar{D}'$ and $\bar{D}''$ respectively, and the function f belongs to the space $L_2(Q)$, then the homogeneous Dirichlet problem (i.e. the boundary-value problem II) has a unique solution such that

$$u \in L_2\Big(0,T; W_2^2(D)\Big) \bigcap \overset{0}{W}{}_2^1(Q), \qquad \sqrt{t}u_{tt} \in L_2(\Omega),$$

$$\sqrt{t}u_{x_i x_j t} \in L_2(Q) \quad \text{for} \quad i,j = 1,\dots m.$$

For the boundary-value problem I, when $\kappa = 1$, the solution may be unstable relative to the entering data.

Proof. For $\kappa = 0$, the existence of the desired solution of boundary-value problem (1.3.69), (1.3.70), (1.3.71$_0$) may be proved easily by the Faedo-Galyerkin method. We suppose that $\{\varphi_k(x')\}$ are the orthonormal eigenfunctions of the homogeneous Dirichlet problem for the operator A in the domain D', and μ_k are the corresponding eigenvalues. We define the approximate solution $\overset{N}{u}(x,t)$ as the finite sum

$$\overset{N}{u}(x,t) = \sum_{k=1}^{N} v_k(x'',t)\varphi_k(x').$$

Acting in a standard way, we obtain the closed system of ordinary differential equations to obtain the functions $v_k(x'',t)$ (x'' is a parameter here). Having the functions $v_k(x'',t)$ and *a priori* estimates, we may pass to the limit for $N \to \infty$ and obtain the desired solution. The necessary *a priori* estimates we may easily obtain by analyzing the equalities

$$\int_0^t\int_{D'} L_0u\cdot u_t dxdt = \int_0^t\int_{D'} fu_t dxdt,$$

$$\int_0^t\int_{D'} L_0u\cdot Au dxdt = \int_0^t\int_{D'} fAu dxdt,$$

$$\int_0^t\int_{D'} L_0u\cdot Au_t dxdt = \int_0^t\int_{D'} fAu dxdt$$

(the index N is omitted here).

Further, the existence of the required solution of the boundary-value problem II for $\kappa = 0$ follows from Theorem 1.3.10.

Now suppose in equation (1.3.69) we have $\kappa = -1$.

To establish the existence of the desired solution of boundary-value problem (1.3.69), (1.3.70), (1.3.71), we use the regularization method. Namely, for $\varepsilon > 0$, we consider the set of equations

$$L_{-1,\varepsilon}u \equiv u_{tt} - \Big(A + \varepsilon C\Big)u_t - Bu - Cu = f(x,t). \tag{1.3.72}$$

The existence of the regular solutions of boundary-value problem (1.3.72), (1.3.70), (1.3.71), may be shown similarly as the existence of the regular solutions of boundary-value problem (1.3.69), (1.3.70), ($1.3.71_0$) for $\kappa = 0$. Moreover, it is easy to show that under the hypotheses of the theorem the equation (1.3.72) may be termwise differentiated with respect to t, and the function $u_{tt}(x,0)$ belongs to the space $L_2(D)$.

The estimates which we need for passing to the limit in equation (1.3.72) may be obtained easily by analyzing the equalities

$$\int_0^t\int_D L_{-1,\varepsilon}u\cdot u_\tau dxd\tau = \int_0^t\int_D fu_\tau dxd\tau,$$

$$\int_0^t\int_D \left(L_{-1,\varepsilon}u\right)_\tau \cdot u_{\tau\tau}dxd\tau = \int_0^t\int_D f_\tau u_\tau dxd\tau,$$

$$\int_0^t\int_D L_{-1,\varepsilon}u \cdot Au_\tau dxd\tau = \int_0^t\int_D fAu_\tau dxd\tau,$$

$$\int_0^t\int_D \left(L_{-1,\varepsilon}u\right)_\tau \cdot Au_{\tau\tau}dxd\tau = \int_0^t\int_D f_\tau Au_{\tau\tau}dxd\tau,$$

$$\int_0^t\int_D \left(L_{-1,\varepsilon}u\right)_\tau \cdot Cu_\tau dxd\tau = \int_0^t\int_D f_\tau Cu_\tau dxd\tau.$$

We note only that the integrals with the products $Au_\tau \cdot Au_\tau$, $Au_\tau \cdot Cu$ and so on, due to the special form of the operators A and C and the domain D are analyzed relatively simply. This is so, because the boundary integrals which arise will vanish (here the (A_0, C_0)-condition holds automatically).

Passing to the limit in a standard way, we obtain the desired solution of boundary-value problem (1.3.69), (1.3.70), (1.3.71) for $\kappa = 0$.

Now we consider possible nonuniqueness in solving the homogeneous Dirichlet problem for equation (1.3.69) for $\kappa = -1$. Suppose we have in this equation that $b^{ij} \equiv b^i \equiv b_0 \equiv 0$; $\{\psi_l(x'')\}$ are the orthonormal eigenfunctions of the homogeneous Dirichlet problem for the operator C in the domain D''; and ν_l are the corresponding eigenvalues. Suppose the indices k and l are such that $\mu_k^2 + 4\nu_l < 0$. For these k and l the function

$$u_{kl}(x,t) = \sin\left(\sqrt{-\frac{\mu_k^2}{4} - \nu_l}\,t\right)\exp\left(\frac{\mu_k}{2}t\right)\varphi_k(x')\psi_l(x'')$$

solves the equation

$$u_{tt} - Au_t - Cu = 0.$$

If the numbers μ_k, ν_l, and T satisfy the relation

$$\frac{\mu_k^2}{4} + \nu_l = -\frac{\pi^2 r^2}{T^2},$$

where r is an integer, then the function $u_{kl}(x,t)$ will vanish everywhere in the boundary of the cylinder Q. This function is an example of a nontrivial solution of the homogeneous Dirichlet problem for equation (1.3.69) for $\kappa = -1$ and a trivial right side.

Finally, we shall establish the third part of the theorem.

We suppose that in equation (1.3.69) $\kappa = 1$. We shall prove that the homogeneous Dirichlet problem (i.e. the boundary-value problem II) for this equation has the desired solution. We shall use again the Galyerkin method. Namely, we define the approximate solutions of the problem in question as

$$u^N(x,t) = \sum_{l=1}^{N} w_l(x',t)\psi_l(x'').$$

The functions $w_l(x',t)$ we choose as solutions of the problem

$$w_{ltt} - Aw_{lt} - Bw_l + \nu_l w_l = f_l(x',t),$$

$$w_l\Big|_{\partial\Omega} = 0,$$

where the functions $f_l(x',t)$ are defined as follows:

$$f_l(x',t) = \int_{D''} f\psi_l(x'')dx''.$$

Existence of the functions $w_l(x',t)$ which have all the required derivatives follows from Theorem 1.3.10. The a *priori* estimates for the set of approximate solutions are obtained in a similar manner to that used for the a *priori* estimates when we proved Theorem 1.3.10. Namely, we analyze the equalities

$$\int_Q Lu \cdot u \exp\Big(\lambda t\Big)dxdt = \int_Q f \cdot u \exp\Big(\lambda t\Big)dxdt,$$

$$\int\limits_Q Lu \cdot Au \exp\Big(\lambda t\Big) dxdt = \int\limits_Q f \cdot Au \exp\Big(\lambda t\Big) dxdt,$$

$$\int\limits_Q Lu \cdot tu_{tt} dxdt = \int\limits_Q f \cdot tu_{tt} dxdt,$$

$$\int\limits_Q Lu \cdot tAu_t dxdt = \int\limits_Q f \cdot tAu_t dxdt,$$

$$\int\limits_Q Lu \cdot Cudxdt = \int\limits_Q f \cdot tCudxdt;$$

(the index N is omitted here; the number λ is an appropriate positive number). Note that the integrals which contain the products $Au{\cdot}Cu$ and $Au_t{\cdot}Cu$, as earlier, are transformed simply, since the operators A, C and the domain D have a special form.

These estimates are sufficient to pass to the limit. The limit function will solve the boundary-value problem II for equation (1.3.69) for $\kappa = 1$.

Instability of solutions of the boundary-value problem I in the case of equation (1.3.69) with $\kappa = 1$ characterizes the following example. The solution of the problem

$$u_{tt} - Au_t + Cu = 0,$$

$$u(x,0) = 0, \qquad u_t(x,0) = \frac{1}{l}\varphi_k(x')\psi_l(x''),$$

$$u\Big|_{\partial D\times(0,T)} = 0$$

is the function

$$u^{kl}(x,t) = \frac{1}{l(z_1 - z_2)}\Big[\exp(z_1 t) - \exp(z_2 t)\Big]\varphi_k(x')\psi_l(x''),$$

where the numbers z_1 and z_2 are defined as follows:

$$z_{1,2} = \frac{\mu_k}{2} \pm \sqrt{\frac{\mu_k^2}{4} - \nu_l}.$$

Evidently, in the norm of the space $L_2(Q)$, we have $u^{kl}(x,t) \to \infty$ for $l \to \infty$; in that time, in the norm of $L_2(D)$ we have

$$u_t^{kl}(x,0) \to 0 \quad \text{for} \quad l \to \infty.$$

The theorem is proved.

From Theorem 1.3.11, in particular, it follows that there exist the equations of composite type of the third order for which both the hyperbolic boundary-value problem and the elliptic boundary-value problem are correct. Of course, we may say that if in equation (1.3.1) the operator A is elliptic and if the boundary Γ of the domain D coincides either with the set σ_1 or with the set σ_0 then the boundary-value problems I and II will coincide and, thus, will both be correct. Theorem 1.3.11 gives the example of an equation whose supports of the boundary data of the boundary-value problems I and II are different. Nevertheless, as indicated above, both these problems will be correct.

1.3.8. Remark on the third order equations of variable direction

One of the classes of third order equations with variable direction of parabolicity was investigated earlier in point 3.3 (equations (1.3.23)). It was studied on the basis of the hyperbolic approach. Here we shall give another example of an equation also with variable direction of parabolicity. To study this equation, we shall apply another method connected with using the solution properties of the boundary-value problem II.

In the cylinder $Q = D \times (0,T)$, $D \subset R^n$, we consider the equation

$$Lu \equiv u_{tt} - K(t)Au_t - Bu = f(x,t), \tag{1.3.73}$$

where the operators A and B are differential operators of the second order of the same form as in Theorem 1.3.11 (in particular, the operator A is assumed to be elliptic in the set $\bar{D}$). The function $K(t)$, except for the smoothness properties, is as follows:

$$\left(T-2t\right)K(t)>0 \quad \text{for} \quad t\neq \frac{T}{2}, \quad K\left(\frac{T}{2}\right)=0.$$

Suppose that Q^+ is the cylinder $D\times(0,\frac{T}{2})$, Q^- is the cylinder $D\times(\frac{T}{2},T)$. According to Theorem 1.3.11, equation (1.3.73) in the cylinder Q^+ has a regular solution satisfying the conditions

$$u(x,0)=0, \quad u_t(x,0)=0, \quad u\Big|_{\partial D\times(0,\frac{T}{2})}=0 \tag{1.3.74}$$

(only trivial initial data are taken for simplicity). We denote this solution by $u^+(x,t)$. Further, again from Theorem 1.3.11, it follows that in the cylinder Q^{-1} equation (1.3.73) has a regular solution satisfying the conditions

$$u(x,T)=0, \quad u_t(x,T)=0, \quad u\Big|_{\partial D\times(\frac{T}{2},T)}=0. \tag{1.3.75}$$

We denote this equation by $u^-(x,t)$. Constructing the function

$$u(x,t)=\begin{cases} u^+(x,t) \text{ for } x\in D,\, t\in\left(0,\frac{T}{2}\right), \\ \\ u^-(x,t) \text{ for } x\in D,\, t\in\left(\frac{T}{2},T\right), \end{cases}$$

we obtain the essential requirement of existence of the regular solution of equation (1.3.73) in the whole cylinder Q: the requirement of continuity of the function $u(x,t)$ and its derivatives in the line $t=\frac{T}{2}$. These conditions may be written in terms of the expressions $(L^+)^{-1}f$, $(L^-)^{-1}f$ (L^+,L^- are the operator L in the cylinders Q^+ and Q^- with conditions (1.3.74) and (1.3.75) respectively) and will be considered as additional conditions imposed on the function $f(x,t)$.

The approach suggested above is based on the hyperbolic approach, i.e. on using the properties of solutions of the boundary-value problem I. As will be shown below, we may apply the other approach to equations (1.3.73) in some cases. This approach is based on using the properties of solutions of the boundary-value problem II; in this case, the solvability will be proved without algebraic conditions imposed on the function $f(x,t)$.

We suppose that in equation (1.3.73) the operator B has the form $B=\beta A$, $\beta=\text{const}>0$. We consider the following boundary-value problem: *find a solution of the equation*

$$Lu \equiv u_{tt} - K(t)Au_t - \beta Au = f(x,t) \tag{1.3.76}$$

satisfying the conditions

$$u(x,0) = u_t(x,0) = u_t(x,T), \qquad u\Big|_{\partial D\times(0,T)} = 0. \tag{1.3.77}$$

Suppose that the function $f(x,t)$ is sufficiently smooth. In the cylinder Q^+, we construct the function $u^+(x,t)$ which solves equation (1.3.76) and satisfies conditions (1.3.74). Subsequently analyzing the equalities

$$\int_0^t\int_D Lu^+ \cdot u_\tau^+ dx d\tau = \int_0^t\int_D f \cdot u_\tau^+ dx d\tau,$$

$$\int_0^t\int_D Lu^+ \cdot Au_\tau^+ dx d\tau = \int_0^t\int_D f \cdot Au_\tau^+ dx d\tau,$$

$$\int_0^t\int_D Lu^+ \cdot A^2u_\tau^+ dx d\tau = \int_0^t\int_D fA^2u_\tau^+ dx d\tau,$$

where $0 < t \leq \frac{T}{2}$ (we suppose additionally that the function $f(x,t)$ vanishes for $x \in \Gamma$), we obtain the estimates

$$\int_D \left[u_t^{+^2}(x,t) + \sum_{i=1}^n u_{x_i}^{+^2}(x,t)\right] dx + \sum_{i=1}^n \int_0^t\int_D K(\tau)u_{x_i\tau}^{+^2}(x,\tau)dx d\tau \leq N_1,$$

$$\int_D \left[\sum_{i=1}^n u_{x_i\tau}^{+^2}(x,t) + \sum_{i,j=1}^n u_{x_ix_j}^{+^2}(x,t)\right] dx + \sum_{i,j=1}^n \int_0^t\int_D K(\tau)u_{x_ix_j\tau}^{+^2}(x,\tau)dx d\tau \leq N_2,$$

$$\int_D \left[\sum_{i,j=1}^n u_{x_ix_jt}^{+^2}(x,t) + \sum_{i,j,k=1}^n u_{x_ix_jx_k}^{+^2}(x,t)\right] dx$$

$$+\sum_{i,j,k=1}^{n}\int_0^t\int_D K(\tau)u^{+^2}_{x_ix_jx_k}(x,\tau)dxd\tau \le N_3.$$

Here the constants $N_1 - N_3$ depend only on the function $f(x,t)$, the domain D, the values T, β, and the coefficients of the operator A. The following estimate will also hold

$$\int_D u^{+^2}_{tt}(x,t)dx \le N_4$$

with the constant N_4 dependent on the same values as constants $N_1 - N_3$.

We denote $w^+(x,t) = Au^+(x,t)$. This function satisfies the equation

$$w^+_{tt} - Aw^+_t - \beta Aw^+ = Af$$

in the cylinder Q^+. Assuming that the function $g = Af$ has the same properties as the function f, we obtain the same properties for the function $w^+(x,t)$ as for the function $u^+(x,t)$. Hence it will follow that the functions u^+, u^+_t, u^+_{tt} have uniquely determined traces for $t = T/2$. We introduce the following notations

$$u^+\left(x,\frac{T}{2}\right) = \varphi_0(x), \quad u^+_t\left(x,\frac{T}{2}\right) = \varphi_1(x), \quad u^+_{tt}\left(x,\frac{T}{2}\right) = \varphi_2(x).$$

For these functions the following inclusions hold

$$\varphi_0(x) \in W_2^5(D), \quad \varphi_1(x) \in W_2^4(D), \quad \varphi_2(x) \in W_2^2(D).$$

Besides, these functions will vanish in Γ and

$$\varphi_2(x) = f\left(x,\frac{T}{2}\right) + \beta A\varphi_0.$$

In the cylinder Q^-, we consider the equation

$$v_{tt} - K(t)Av_t - \left(\beta + K_t\right)Av = f_t \tag{1.3.78}$$

obtained after formal differentiation of equation (1.3.76) with respect to t. We supply equation (1.3.78) with conditions

$$v_t\left(x,\frac{T}{2}\right)=\varphi_2(x),\quad v(x,T)=0,\quad v\Big|_{\Gamma\times(\frac{T}{2},T)}=0. \tag{1.3.79}$$

For convenience, we introduce the function

$$\tilde{w}=v-\frac{2}{T}\left(\frac{T}{2}-t\right)\left(t-T\right)\varphi_2(x).$$

This function satisfies equation (1.3.78) with the changed right side and satisfies homogeneous conditions (1.3.79).

We suppose that for t such that $\frac{T}{2}\le t\le T$, the following conditions

$$\beta+K_t\le 0,\qquad \beta+\frac{1}{2}K_t\le 0$$

hold. Then the function $\tilde{w}(x,t)$ may be uniquely determined from equation (1.3.78) (with the changed right side) and homogeneous boundary-value conditions (1.3.79). This may be proved similarly as we have established the existence of solution of the boundary-value problem II in the third part of Theorem 1.3.11, i.e. by the Galyerkin method. Subsequently multiplying the equation for the function $\tilde{w}(x,t)$ from the right and from the left by the functions $\tilde{w}$, $A\tilde{w}$, $A^2\tilde{w}$ and integrating the equalities obtained from the right and from the left over the cylinder Q^-, we obtain the estimate

$$\int\limits_{Q^-}\left[\tilde{w}_{tt}^2+\sum_{i=1}^{n}w_{x_it}^2+\sum_{i,j=1}^{n}\tilde{w}_{x_ix_j}^2+\sum_{i,j=1}^{n}\tilde{w}_{x_ix_jt}^2\right]dxdt\le N_5.$$

Here the constant N_5 depends only on the function $f(x,t)$, the domain D, the number T, and the coefficients of the operator A. The analogous estimate, evidently, will hold and for the function $v(x,t)$.

We set

$$u^-(x,t)=\int\limits_{T/2}^{t}v(x,\tau)d\tau+\varphi_0(x),$$

$$u(x,t)=\begin{cases} u^+(x,t) \text{ for } x\in D,\ t\in\left[0,\frac{T}{2}\right], \\ u^-(x,t) \text{ for } x\in D, t\in\left[\frac{T}{2},T\right], \end{cases}$$

The function $u(x,t)$ will be the regular solution of problem (1.3.76), (1.3.77). Really, integrating equation (1.3.78) with respect to t from $T/2$ to the arbitrary point, we obtain

$$u_{tt}^- - \varphi_2(x) - K(t)Au_t^- + K\left(\frac{T}{2}\right)Av\left(x,\frac{T}{2}\right) - \beta Au^-$$

$$+\beta A\varphi_0 = f(x,t) - f\left(x,\frac{T}{2}\right).$$

The estimate for the function v gives the existence of the trace of the function Av for $t = T/2$; therefore, $K(T/2)Av(x,T/2) = 0$. Taking account of the connection between the functions $\varphi_0(x)$, $\varphi_2(x)$, and $f(x,T/2)$, we see that equation (1.3.76) holds in the cylinder Q^-. The existence of all the derivatives entering the equation (1.3.76) and their membership in the space $L_2(Q)$ is evident. Thus, the regular solution of problem (1.3.76), (1.3.77) is constructed.

Conditions

$$\beta + K_t \le 0, \quad \beta + \frac{1}{2}K_t \le 0 \quad \text{for} \quad t\in\left(\frac{T}{2},T\right)$$

hold, for example, for equation

$$u_{tt} - (T-2t)u_{xxt} - u_{xx} = f(x,t).$$

Suppose now that for the function $K(t)$ the following conditions

$$\beta + mK_t(t) > 0 \quad \text{for} \quad t\in\left[0,\frac{T}{2}\right], \qquad m = 1,\dots,m_0-1,$$

$$\beta + m_0K_t(t) \le 0 \quad \text{for} \quad t\in\left[\frac{T}{2},T\right]$$

hold, where m_0 is a certain natural number. Then, if other conditions hold, we may construct the regular solution of equation (1.3.76) with the help of the boundary conditions

$$u(x,0) = u_t(x,0) = 0, \quad \frac{\partial^{m_0} u}{\partial t^{m_0}} = 0, \quad u\Big|_{\Gamma\times(0,T)} = 0.$$

Really, if we differentiate equation (1.3.76) with respect to t $(m_0 - 1)$ times in the cylinder Q^+, it conserves the hyperbolic properties; and, after differentiating one time in the cylinder Q^-, we obtain the elliptic properties.

1.3.9. Nonlocal problems for equations of composite type

Boundary-value problems I, I′, and II stated above are the so-called local problems, i.e. problems with point conditions. If local boundary conditions (one or more) are changed by nonlocal conditions; i.e. by such conditions where a connection between boundary operator values in various boundary points is given, then such problems are called nonlocal problems. Some of these problems will be considered in this section.

In a bounded domain D of the space R^n with a boundary Γ we consider the equation

$$Lu \equiv \left(\frac{\partial}{\partial x_1} + \mu(x)\right) \Delta u + c(x)u = f(x), \tag{1.3.80}$$

where

$$x = (x_1, x_2, \ldots, x_n) = (x_1, x'), \qquad x' = (x_2, \ldots, x_n).$$

Let a mapping g transform a set σ_1 on a set σ_2 and suppose this mapping is one-to-one.

The boundary-value problem: find a solution of equation (1.3.80) satisfying the conditions

$$u(x)\Big|_{x\in\sigma_1} = h(\eta)u(\eta)\Big|_{\eta\in\sigma_2}, \tag{1.3.81}$$

$$u_{x_1} + \mu u\Big|_{\Gamma} = 0, \tag{1.3.82}$$

where the point $\eta \in \sigma_2$ is the image of the point $x \in \sigma_1$ under the mapping g; $h(\eta)$ is a given function.

Suppose that the sets σ_1 and σ_2 are defined as follows:

$$\sigma_1 = \Big\{x = (x_1, x') : \, x_1 = F_1(x'), \quad x' \in G\Big\},$$

$$\sigma_2 = \Big\{x = (x_1, x') : \, x_1 = F_2(x'), \quad x' \in G\Big\},$$

where G is a bounded domain of the space R^n; and D is the set

$$D = \Big\{x = (x_1, x') : \, F_1(x') < x_1 < F_2(x'), \quad x' \in G\Big\}.$$

Note that the mapping g induces the mapping $\Phi : G \to G$ such that if $x = (x_1, x')$ is a point of the set σ_1 and $\eta = (\eta_1, \eta')$, $\eta' = (\eta_2, \dots, \eta_n)$ is the image of x under the mapping g, then we have the equality $x' = \Phi(\eta')$.

We introduce now the following notations

$$k_0 = \max_{i=2,\dots,n} \max_{\bar{D}} \Big|\mu_{x_i}(x)\Big|,$$

$$c_0(x) = \mu(x)\Big[\mu_{x_1 x_1}(x) - c(x) - \Delta_{n-1}\mu(x)\Big],$$

$$c_1(x) = 2\mu_{x_1}(x)\mu(x) + \mu_{x_1 x_1}(x) - c(x) - \Delta_n \mu(x),$$

where Δ_{n-1} is the Laplace operator with respect to the variables $x_2, \dots, x_n$. Let m_0 be such a constant that for each function $v(x)$ from the space $\overset{0}{W}{}_2^1(D)$ the inequality

$$\int_D v^2 dx \le m_0 \sum_{i=1}^{n} \int_D v_{x_i}^2 dx$$

holds. Finally, suppose that $\alpha(\lambda)$, for a real λ, is maximal value of the function $\exp(\lambda x_1)$ determined for those real x_1 for which there exists a point x' from the space R^{n-1} such that (x_1, x') is a point from the set $\bar{D}$.

Theorem 1.3.12. *Let a mapping $\Phi(\eta')$ be a diffeomorphism which Jacobian $J(\eta')$ does not vanish in the set G and which may be continuously extended in ∂G. Suppose also that the conditions below hold:*

$$\mu_{x_1}(x) \geq 0, \qquad c_0(x) - \frac{1}{2}c_{1x_1}(x) \geq \bar{c}_0 > 0$$

for all x from $\bar{D}$, and

$$c_1\Big(F_2(\eta'),\eta'\Big) - c_1\Big(F_2(\Phi(\eta')),\Phi(\eta')\Big)h^2\Big(F_2(\eta'),\eta'\Big)\Big|J(\eta')\Big| \geq 0$$

for all η from G, and there exists a constant λ_0 such that, for $x \in \bar{D}$,

$$\left(\mu(x) - \frac{\lambda_0}{2}\right)\exp\Big(\lambda_0 x_1\Big) \geq \mu_0 > 0;$$

and for $\eta \in G$,

$$\exp\Big(\lambda_0 F_2(\eta')\Big) - h^2\Big(F_2(\eta'),\eta'\Big)\exp\Big(\lambda_0 F_1(\Phi(\eta'))\Big)\Big|J(\eta')\Big| \geq 0.$$

Finally, we suppose that the constants $\bar{c}_0, \lambda_0$, and μ_0 are such that

$$\bar{c}_0 + \frac{1-\delta^2}{m_0}\cdot\frac{\mu_0^2}{\alpha^2(\lambda_0)} - \frac{k_0^2}{\delta_0^2} > 0$$

for a certain number δ_0 from the interval $(0,1)$.

Then the regular solution of boundary-value problem (1.3.80)-(1.3.82) is unique.

Proof. We suppose that in equation (1.3.80) $f \equiv 0$. Then, the following equality

$$-\int_D Lu\Big(u_{x_1} + \mu u\Big)dx = 0$$

holds. Integrating by parts in the left side of this equality, we obtain the following equality

$$\int_D \left\{ \sum_{i=1}^n \left[\frac{\partial}{\partial x_i} \Big(u_{x_1} + \mu u \Big) \right]^2 + 2\mu_{x_1} u_{x_1}^2 + \left(c_0 - \frac{1}{2} c_{1x_x} \right) u^2 \right\} dx dy$$

$$-2 \sum_{i=2}^n \int_Q \mu_{x_i} \cdot u \cdot \frac{\partial}{\partial x_i} \Big(u_{x_1} + \mu u \Big) dx - \frac{1}{2} \int_\Gamma c_1 u^2 \nu_1 d\gamma = 0. \qquad (1.3.83)$$

The boundary integral

$$I = -\frac{1}{2} \int_\Gamma c_1 u^2 \nu_1 d\gamma$$

from the left side of (1.3.83) may be transformed as follows:

$$I = -\frac{1}{2} \int_\Gamma c_1 \Big(F_1(x'), x' \Big) u^2 \Big(F_1(x'), x' \Big) dx'$$

$$+\frac{1}{2} \int_\Gamma c_1 \Big(F_2(\eta'), \eta' \Big) u^2 \Big(F_2(\eta'), \eta' \Big) d\eta'.$$

In the first of these integrals, we make the change $x' = \Phi(\eta')$.

Note that boundary condition (1.3.81) may be written in the form

$$u\Big(F_1(\Phi(\eta')), \Phi(\eta') \Big) = h\Big(F_2(\eta'), \eta' \Big) u\Big(F_2(\eta'), \eta' \Big).$$

Taking this into account we rewrite the boundary integral

$$I = \frac{1}{2} \int_G c_1 \Big(F_2(\eta'), \eta' \Big) - c_1 \Big(F_1(\Phi(\eta')), \Phi(\eta') \Big) h^2 \Big(F_2(x'), x' \Big) \Big| J(\eta') \Big| d\eta'.$$

By the hypotheses of the theorem, we obtain that I is nonnegative.

Applying the Young inequality, the hypotheses of the theorem and nonnegativeness of the integral I to equality (1.3.83), we obtain

$$\int\limits_D \left[\frac{\partial}{\partial x_1}\left(u_{x_1} + \mu u\right)\right]^2 dx + \left(1 - \delta_0^2\right) \sum_{i=2}^{n} \int\limits_D \left[\frac{\partial}{\partial x_i}\left(u_{x_1} + \mu u\right)\right]^2 dx$$

$$+\bar{c}_0 \int\limits_D u^2 dx \leq \frac{k_0^2}{\delta_0^2} \int\limits_D u^2 dx. \tag{1.3.84}$$

We consider now the equation

$$u_{x_1} + \mu u = v.$$

Integrating by parts in the equality

$$\int\limits_D \left(u_{x_1} + \mu u\right) u \exp\left(\lambda_0 x_1\right) dx = \int\limits_D v u \exp\left(\lambda_0 x_1\right) dx,$$

and taking account of the condition $v = 0$ in Γ and the theorem conditions, we obtain the inequality

$$\int\limits_D u^2 dx \leq \frac{\alpha^2(\lambda_0)}{\mu_0^2} \int\limits_D v^2 dx.$$

Combining this inequality and inequality (1.3.84) and taking into account the condition for the constants $\bar{c}_0$, δ_0, m_0, μ_0, $\alpha(\lambda_0)$, and k_0, we see that

$$\int\limits_D u^2 dx \leq 0.$$

This inequality yields $u \equiv 0$ in the domain D. The theorem is proved.

Existence of the solution of the boundary-value problem we shall establish under the condition that the function μ depends only on x_1. This assumption is made only to simplify the proof.

We suppose that $\eta' = \Psi(x')$ is inverse to $\Phi(\eta')$. Denote by $h_0(x')$ the function

$$h_0(x') = h\Big(F_2(\Psi(x')), \Psi(x')\Big).$$

Theorem 1.3.13. *Let all the conditions of Theorem 1.3.12 hold and suppose that, for each η' from the set $\bar{G}$, the inequality*

$$h^2\Big(F_2(\eta'),\eta'\Big)\cdot\Big|J(\eta')\Big| \le k < 1$$

holds. If a function $f(x)$ belongs to the space $L_2(D)$, then problem (1.3.80)-(1.3.82) has a solution $u(x)$ such that the following inclusions hold

$$u \in L_2(D), \quad u_{x_1} \in L_2(D), \quad u_{x_1} + \mu u \in W_2^2(D).$$

Proof. We apply the method of continuation by parameter. For this purpose, we consider the set of operators

$$L_\tau u \equiv \Delta\Big(u_{x_1} + \mu u\Big) + \tau\Big[-2\mu'\cdot u_{x_1} + \Big(c - \mu''\Big)u\Big],$$

where $\tau \in [0,1]$ and the set of corresponding boundary-value problems: *find a solution of the equation*

$$L_\tau u = f(x), \tag{1.3.80$_\tau$}$$

satisfying conditions (1.3.81) and (1.3.82).

When $\tau = 0$, for the function $v = u_{x_1} + \mu u$, we have the usual Dirichlet problem for the Poisson equation. From the well-known results on solvability of boundary-value problems for elliptic equations (Ladyzhenskaya and Ural'tseva, 1973), it follows that the function $v(x)$ may be uniquely determined; and this function will belong to the space $W_2^2(D)$. We shall show that, if we know the function $v(x)$, we may find the function $u(x)$. Namely, the function $u(x)$ we define from the equation

$$u_{x_1} + \mu(x_1)u = v \tag{1.3.85}$$

and boundary condition (1.3.81). The solution of this equation is as follows:

$$u(x) = \exp\left(-\int\limits_{F_1(x')}^{x_1}\mu(z)dz\right)\cdot\int\limits_{F_1(x')}^{x_1} v(z,x')\exp\left(\int\limits_{F_1(x')}^{z}\mu(t)dt\right)dz + w(x'),$$

where the function $w(x')$ is a certain as yet unknown function. In this representation of the function $u(x)$, we set $x_1 = F_1(x')$. Hence we obtain the equality

$$u\Big(F_1(x'), x'\Big) = w(x').$$

Now, in the representation of the function $u(x)$, we change the variable x' by the variable η' and set $x_1 = F_2(\eta)$. As a result, we obtain the equality

$$u\Big(F_2(\eta'), \eta'\Big) = v_0(\eta') + w(\eta'),$$

where the function $v_0(\eta')$ is the known function. Taking into account boundary condition (1.3.81), we obtain

$$w(x') = h\Big(F_2(\eta'), \eta'\Big)v_0(\eta') + h\Big(F_2(\eta'), \eta'\Big)w(\eta').$$

In other words, the function $w(x')$ must solve the functional equation

$$w(x') = h_0(x')v_0\Big(\Psi(x')\Big) + h_0(x')w\Big(\Psi(x')\Big). \tag{1.3.86}$$

Similar equations are studied in the monograph by Kuczma (1968), where the general theorems on solvability of such (and even more general) equations are formulated in the classes of continuous functions. The conditions pointed out in the theorem formulation provide solvability of equation (1.3.86) in the space $L_2(G)$. Now we shall prove this.

We make the notations

$$v_1(x') = h_0(x')v_0\Big(\Psi(x')\Big),$$

$$(Aw)(x') = v_1(x') + h_0(x')w\Big(\Psi(x')\Big).$$

The following relations hold

$$\Big\|Aw_1 - Aw_2\Big\|_{L_2(G)} = \Big\|h_0(x')\Big[w_1\Big(\Psi(x')\Big) - w_2\Big(\Psi(x')\Big)\Big]\Big\|_{L_2(G)}$$

$$= \left\{ \int\limits_G h_0^2(x') \Big[w_1\Big(\Psi(x')\Big) - w_2\Big(\Psi(x')\Big) \Big]^2 dx' \right\}^{\frac{1}{2}}$$

$$= \left\{ \int\limits_G h^2\Big(F_2(\eta'), \eta'\Big) \Big[w_1(\eta') - w_2(\eta') \Big]^2 \Big| J(\eta') \Big| d\eta' \right\}^{\frac{1}{2}}$$

$$\leq k \Big\| w_1 - w_2 \Big\|_{L_2(G)}.$$

Therefore, the operator A is a contractive operator in the space $L_2(G)$. From the classical theorem of functional analysis it follows that the equation

$$Aw = w$$

is solvable in the space $L_2(G)$. In other words, the function $w(x')$, as solution of equation (1.3.86), is correctly determined; and the function $w(x')$ will be the element of the space $L_2(G)$.

The function $w(x')$ allows us to obtain the function $u(x)$, solution of problem (1.3.85), (1.3.81); and the function $u(x)$ will be such that its derivative u_{x_1} will belong to the space $L_2(D)$. As the function $u_{x_1} + \mu u$ belongs to the space $W_2^2(D)$, the function $u(x)$ will solve boundary-value problem (1.3.80$_0$), (1.3.81), (1.3.82).

Now let τ be an arbitrary number from $[0,1]$. Repeating the considerations of Theorem 1.3.12, we may easily show that, for solutions of equation (1.3.80$_\tau$) satisfying conditions (1.3.81), (1.3.82), the following estimate will hold uniform with respect to τ

$$\Big\| u_{x_1} + \mu u \Big\|_{W_2^1(D)} + \Big\| u_{x_1} \Big\|_{L_2(D)} + \Big\| u \Big\|_{L_2(D)} \leq M_1.$$

The constant M_1, in this estimate, depends only on the norm of $f(x)$ in the space $L_2(D)$ and on the coefficients of the operator L. Further, the second basic inequality for elliptic operators gives the estimate

$$\Big\| u_{x_1} + \mu u \Big\|_{W_2^2(D)} \leq M_2$$

with the estimate M_2 dependent on the same values as the constant M_1.

These estimates, uniform with respect to τ, allow us to realize the scheme of the method of continuation by parameter and, thus, to establish solvability of problem (1.3.80$_\tau$), (1.3.81), (1.3.82) in the desired class for all τ (in particular, for $\tau = 1$). The theorem is proved.

Note that if the conditions of Theorem 1.3.12 fail, it may be that boundary-value problem (1.3.80), (1.3.82) will be ill-posed. We now show this.

Let a domain D be a cylindrical domain, $D = (0,1) \times G$, where G is a bounded domain from the space R^{n-1} of the variables $x_2, \dots, x_n$ and suppose the boundary of G is a smooth manifold.

The simplest example of an ill-posed problem of type (1.3.80)-(1.3.82) is the problem

$$\frac{\partial}{\partial x_1} \Delta x = 0,$$

$$u(0, x') = u(1, x'),$$

$$u_{x_1}\Big|_{\Gamma} = 0.$$

This problem is solved by the function which is equal to a constant.

A more complicated example is the problem

$$\frac{\partial}{\partial x_1} \Delta u + \lambda u = 0, \tag{1.3.87}$$

$$u(0, y) = \beta u(1, y), \tag{1.3.88}$$

$$u_{x_1}\Big|_{\Gamma} = 0. \tag{1.3.89}$$

We suppose that in equation (1.3.87) the number λ is a fixed negative number, and the number β is a varying parameter. We restrict ourselves to the case $\beta \geq 0$. If $0 \leq \beta < 1$, then boundary-value problem (1.3.87)-(1.3.89) will have a unique solution $u(x) \equiv 0$. Suppose that $\beta \geq 1$. We shall seek a particular solution of problem (1.3.87)-(1.3.89) as a function

$$u_k(x) = c_k(x_1) v_k(x'),$$

where $v_k(x')$ is an eigenfunction of the homogeneous Dirichlet problem for the $(n-1)$-dimensional Laplace operator in the domain G:

$$\Delta_{n-1}v_k = z_k v_k, \qquad v_k\Big|_{\partial G} = 0$$

(z_k is the corresponding eigenvalue). In order that the function $u_k(x)$ solves problem (1.3.87)–(1.3.89), it is necessary that the function $c_k(x_1)$ satisfies the equation

$$c_k'''(x_1) + z_k c_k'(x_1) + \lambda c_k(x_1) = 0$$

and the conditions

$$c_k(0) = c_k(1), \qquad c_k'(0) = \beta c_k'(1) = 0.$$

The characteristic equation of this differential equation is as follows:

$$t^3 + z_k t + \lambda = 0.$$

If the condition

$$\frac{\lambda^2}{4} + \frac{z_k^3}{27} < 0 \tag{1.3.90}$$

holds, then the roots of this equation will be real and distinct. As $z_k < 0$ for all k and $z_k \to -\infty$ for $k \to \infty$ then, for k sufficiently large, condition (1.3.90) will hold. Further, taking an appropriate k, we may satisfy the inequality

$$-z_k + \lambda > 0. \tag{1.3.91}$$

In this case, two roots of the characteristic equation will be negative and one will be positive. We denote them by $t_{1,k}$, $t_{2,k}$, and $t_{3,k}$ so that $t_{1,k} < t_{2,k} < 0 < t_{3,k}$. Note also that $t_{1,k} + t_{2,k} + t_{3,k} = 0$.

The functions $c_k(x_1)$ have the following form:

$$c_k(x_1) = \alpha_{1,k} \exp\Big(t_{1,k}x_1\Big) + \alpha_{2,k} \exp\Big(t_{2,k}x_1\Big) + \alpha_{3,k} \exp\Big(t_{3,k}x_1\Big),$$

where the constants $\alpha_{1,k}$, $\alpha_{2,k}$, and $\alpha_{3,k}$ solve the system

$$\Big[1-\beta\exp\Big(t_{1,k}\Big)\Big]\alpha_{1,k} + \Big[1-\beta\exp\Big(t_{2,k}\Big)\Big]\alpha_{2,k} + \Big[1-\beta\exp\Big(t_{3,k}\Big)\Big]\alpha_{3,k} = 0,$$

$$t_{1,k}\alpha_{1,k} + t_{2,k}\alpha_{2,k} + t_{3,k}\alpha_{3,k} = 0,$$

$$t_{1,k}\exp\Big(t_{1,k}\Big)\alpha_{1,k} + t_{2,k}\exp\Big(t_{2,k}\Big)\alpha_{2,k} + t_{3,k}\exp\Big(t_{3,k}\Big)\alpha_{3,k} = 0.$$

The determinant of this system is $K + \beta N$, where

$$K = t_{1,k}\Big(t_{3,k} - t_{2,k}\Big)\exp\Big(t_{1,k}\Big) + t_{2,k}\Big(t_{1,k} - t_{3,k}\Big)\exp\Big(t_{2,k}\Big)$$

$$+t_{3,k}\Big(t_{2,k} - t_{1,k}\Big)\exp\Big(t_{3,k}\Big),$$

$$N = t_{1,k}\Big(t_{3,k} - t_{2,k}\Big)\exp\Big(-t_{1,k}\Big) + t_{2,k}\Big(t_{1,k} - t_{3,k}\Big)\exp\Big(-t_{2,k}\Big)$$

$$+t_{3,k}\Big(t_{2,k} - t_{1,k}\Big)\exp\Big(-t_{3,k}\Big).$$

As the inequalities

$$t_{2,k}\Big(t_{1,k} - t_{3,k}\Big) > 0, \quad t_{3,k}\Big(t_{2,k} - t_{1,k}\Big) > 0, \quad t_{1,k}\Big(t_{3,k} - t_{2,k}\Big) > 0,$$

$$\exp\Big(t_{1,k}\Big) < \exp\Big(t_{2,k}\Big) < \exp\Big(t_{3,k}\Big)$$

hold, then the number K is positive and the number N is negative. Therefore, the function $\varphi(\beta) = K + \beta N$ is a monotonically decreasing function. Thus, for each natural k which satisfies the conditions (1.3.90) and (1.3.91) there exists only one value $\beta = \beta_k$ for which the determinant of the system for $\alpha_{1,k}$, $\alpha_{2,k}$ and $\alpha_{3,k}$ vanishes. This denotes that this system has nontrivial solutions, which, in its turn, provides the existence of nontrivial solutions of boundary-value problem (1.3.87)-(1.3.89).

These considerations indicate that boundary-value problem (1.3.87)–(1.3.89) for $\beta \geq 1$ may be ill-posed. Moreover, we have obtained that for each fixed negative λ there exist a countable set $\{\beta_k\}$ of the values of the parameter β ($\beta_k \geq 1$ and $k \geq k_0$, where k_0 is a minimal natural number which satisfies conditions (1.3.90) and (1.3.91)) such that boundary-value problem (1.3.87)-(1.3.89) has a nontrivial solution for such β_k. In other words, the parameter β in boundary-value problem (1.3.87)-(1.3.89) may be considered as a special parameter.

Remark. If in equation (1.3.87) the number λ is positive, then the change $t = 1 - x_1$ reduces problem (1.3.87)-(1.3.89) to the problem

$$\frac{\partial}{\partial t}\Delta v - \lambda v = 0,$$

$$v(0, x') = \frac{1}{\beta} v(1, x'),$$

$$v_t\Big|_{\partial D} = 0.$$

Thus, problem (1.3.87)-(1.3.89) will be well-posed for conditions $\lambda > 0$, $|\beta| > 1$ also.

We shall give now some examples of well-posed nonlocal problems for equations of composite type. We restrict ourselves to model cases.

In the cylinder $Q = D \times (0, T)$, $D \subset R^n$ we consider the equation

$$Lu \equiv u_{tt} - Au_t - Bu = f(x,t), \tag{1.3.92}$$

where A and B are second order operators of the form

$$Au = \frac{\partial}{\partial x_i}\left(a^{ij}(x)u_{x_j}\right) - a_0(x)u, \quad Bu = \frac{\partial}{\partial x_i}\left(b^{ij}(x)u_{x_j}\right) - b_0(x)u.$$

The operator A is assumed to be elliptic in the set $\bar{D}$ and the operator B is elliptic-parabolic in the set $\bar{D}$.

We consider the boundary-value problem: *find solutions of equation (1.3.92) satisfying the conditions*

$$u(x,0) = \int_0^T h(t)u(x,t)dt + u_0(x), \tag{1.3.93}$$

$$u_t(x,0) = u_1(x), \tag{1.3.94}$$

$$u\Big|_{\partial D\times(0,T)} = 0, \tag{1.3.95}$$

where the functions $h(t)$, $u_0(x)$, and $u_1(x)$ are given.

If the function $h(t)$ is equal to zero identically, then boundary-value problem (1.3.92)-(1.3.95) becomes the usual composite initial boundary-value problem studied before. If we suppose that the function $h(t)$ coincides with the Dirac function concentrated at the point T, then boundary-value problem (1.3.92)-(1.3.95) becomes a half-periodic boundary-value problem.

Let the elliptic-parabolic conditions for the operator B hold in the form (1.3.25) and suppose that β_0 is the constant

$$\beta_0 = \max(1, \lambda_1)$$

(λ_1 is a constant from condition (1.3.25)).

Theorem 1.3.14. *Let operators A and B be symmetric in $\bar{D}$, the operator A be elliptic in $\bar{D}$ and the operator B be elliptic-parabolic in $\bar{D}$. Suppose also that the function $h(t)$ is continuous in the interval $[0,T]$ and*

$$a_0(x) \geq 0, \quad b_0(x) \geq 0 \quad \textit{for} \quad x \in \bar{D};$$

$$\beta_0 \cdot T \cdot \exp(1) \int_0^T h^2(t)dt < 1;$$

$$\int_0^T h(t)dt \neq 1.$$

If the function $f(x,t)$ belongs to the space $L_2(D)$, the function $u_0(x)$ belongs to the space $W_2^2(D) \cap \overset{0}{W}{}_2^1(D)$, and the function $u_1(x)$ belongs to the space $\overset{0}{W}{}_2^1(D)$, then boundary-value problem (1.3.92)-(1.3.95) has a solution $u(x,t)$ which satisfies the inclusions

$$u \in L_2\Big(0,T; W_2^2(D) \bigcap \overset{0}{W}{}_2^1(D)\Big), \quad u_t \in L_2\Big(0,T; W_2^2(D) \bigcap \overset{0}{W}{}_2^1(D)\Big),$$

$$u_{tt} \in L_2(Q), \quad u_{x_i x_j t} \in L_2(D) \quad \text{for} \quad i,j = 1,\dots,n.$$

Proof. We restrict ourselves to an outline of the proof.
To this end, we consider the one-parameter operator set

$$L_\rho u = (1-\rho)\Big(u_{tt} - Au_t\Big) - \rho B u.$$

For $\rho = 0$, boundary conditions (1.3.94) and (1.3.95) allow us to determine the function $v = u_t$ uniquely as the solution of the equation

$$L_0 u = f(x,t).$$

Condition (1.3.93) allows us to determine the function $u(x,t)$. Really, the solution of equation $u_t = v$ has the form

$$u(x,t) = \int_0^t v(x,\tau)d\tau + w(x). \tag{1.3.96}$$

We multiply this equality from the right and from the left by the function

$h(t)$ and integrate the equality obtained over the interval $[0, T]$. As a result we obtain

$$\int_0^T h(t)u(x,t)dt = v_1(x,t) + w(x)\int_0^T h(t)dt.$$

Boundary condition (1.3.93) denotes that

$$v_t(x,t) + w(x)\int_0^T h(t)dt = w(x).$$

The hypotheses of the theorem and this equality uniquely determine the function $w(x)$. By the function $w(x)$ we determine the function $u(x,t)$.

We show that for solutions of boundary-value problem (1.3.92)-(1.3.95) the satisfactorily *a priori* estimates hold.

Let λ be a negative number. The equality

$$\int_Q Lu \cdot u_t \exp\Big(\lambda t\Big)dxdt = \int_Q f \cdot u_t \exp\Big(\lambda t\Big)dxdt$$

may be easily transformed as

$$-\frac{\lambda}{2}\int_Q u_t^2 \exp\Big(\lambda t\Big)dxdt + \int_Q a^{ij}u_{x_j t}\exp\Big(\lambda t\Big)dxdt$$

$$+\int_Q a_0 u_t^2 \exp\Big(\lambda t\Big)dxdt - \frac{\lambda}{2}\int_Q b^{ij}u_{x_i}u_{x_j}\exp\Big(\lambda t\Big)dxdt$$

$$-\frac{\lambda}{2}\int_Q b_0 u^2 \exp\Big(\lambda t\Big)dxdt + \frac{\exp(T)}{2}\int_D u_t^2(x,T)dx$$

$$+\frac{\exp(T)}{2}\int_D b^{ij}u_{x_i}(x,T)u_{x_j}(x,T)dx + \frac{\exp(T)}{2}\int_D b_0 u^2(x,T)dx$$

$$= \int_Q f \cdot u_t \exp\Big(\lambda t\Big)dxdt + \frac{1}{2}\int_D u_1^2 dx + \frac{1}{2}\int_D b^{ij}u_{x_i}(x,0)u_{x_j}(x,0)dx$$

$$+\frac{1}{2}\int_D b_0 u^2(x,0)dx. \tag{1.3.97}$$

The third term from the right side of equality (1.3.97) may be estimated from above

$$\frac{1}{2}\int_D b^{ij}u_{x_i}(x,0)u_{x_j}(x,0)dx = \frac{1}{2}\int_D b^{ij}\left[\int_0^T h(t)u_{x_i}(x,t)dt + u_{0x_i}\right]$$

$$\times\left[\int_0^T h(t)u_{x_j}(x,t)dt + u_{0x_j}\right]dx$$

$$\le \int_D b^{ij}\left(\int_0^T h(t)u_{x_i}(x,t)dt\right)\left(\int_0^T h(t)u_{x_j}(x,t)dt\right)dx + \int_D b^{ij}u_{0x_i}u_{0x_j}dx$$

$$\le \lambda_1\int_D b^{i,0}\left(\int_0^T h(t)u_{x_i}(x,t)dt\right)^2 dx + \int_D b^{ij}u_{0x_i}u_{0x_j}dx$$

$$\le \lambda_1\int_0^T h^2(t)dt\cdot\int_Q b^{i,0}u_{x_i}^2dxdt + \int_D b^{ij}u_{0x_i}u_{0x_j}dx.$$

Analogously we estimate the fourth term of the right side of equality (1.3.97)

$$\frac{1}{2}\int_D b_0u^2(x,0)dx \le \int_Q b_0u^2dxdt + \int_D b_0u^2dx.$$

For negative λ, we have the inequality

$$-\lambda\exp\Big(\lambda t\Big) \ge -\lambda\exp\Big(\lambda T\Big).$$

The function $\varphi(\lambda) = -\lambda\exp(\lambda T)$ attains the maximum for $\lambda = -T$. Using the hypotheses of the theorem, we obtain the estimate

$$\int_Q \left(u_t^2 + \sum_{i=1}^n u_{x_i t}^2 \right) dxdt \leq N_1, \tag{1.3.98}$$

where the constant N_1 depends only on the coefficients of the operator L, the functions $f(x,t)$, $u_0(x)$, $u_1(x)$, $h(t)$, and on the number T.

Taking account of equality (1.3.96) and estimate (1.3.98), we obtain the estimate

$$\int_Q u^2 dxdt \leq N_2,$$

where the constant N_2 depends on the same values as the constant N_1.

The further estimate we obtain by analyzing the equalities

$$\int_Q Lu \cdot Au \exp\left(\lambda t\right) dxdt = \int_Q f \cdot Au \exp\left(\lambda t\right) dxdt,$$

$$\int_Q Lu \cdot Au_t dxdt = \int_Q f \cdot Au_t dxdt,$$

$$\int_Q Lu \cdot u_{tt} dxdt = \int_Q f \cdot u_{tt} dxdt.$$

In this case, we follow similar considerations as were used above. These estimates correspond to all the inclusions from the theorem. Similar estimates will be valid for solutions of equation

$$L_\rho u = f(x,t)$$

satisfying conditions (1.3.93)-(1.3.95). The constants in the right sides of these estimates will not depend on the parameter ρ. As these estimates will hold for the function $f(x,t)$ from the space $L_2(Q)$, we may apply to the set of the boundary-value problems the method of extension by the parameter. As a result, we obtain the desired solution. The theorem is proved.

We may similarly investigate the boundary-value problem: *find solutions of equation (1.3.92) satisfying the conditions*

$$u(x,0) = u_0(x), \tag{1.3.93'}$$

$$u_t(x,0) = \int_0^T h(t)u_t(x,t)dt + u_1(x) \tag{1.3.94'}$$

and condition (1.3.95). To establish the existence of regular solutions of this boundary-value problem we need to use the results on solvability of nonlocal boundary-value problems for parabolic equations obtained by Shelukhin (1992). The *a priori* estimates are established similar to those obtained above. It is not difficult to investigate the boundary-value problem for equation (1.3.92) when both initial conditions are replaced by nonlocal conditions of type (1.3.93), (1.3.94) and the problem where initial conditions (one or both) are replaced by the conditions

$$u(x,0) = h_0(x)u(x,T) + u_0(x),$$

$$u_t(x,0) = h_1(x)u_t(x,T) + u_1(x).$$

Chapter 2.

Solvability of inverse problems and other applications

2.1. INVERSE PROBLEMS FOR EQUATIONS WITH PARTIAL DERIVATIVES

2.1.1. Linear inverse problems for elliptic and parabolic equations

We denote as a linear inverse problem the problem of determination of a solution of a parabolic or elliptic equation and its right side simultaneously. Similar inverse problems were studied, for example, by Anikonov (1985), Prilepko and Kostin (1992), Lorenzi and Prilepko (1992), Engl *et al.*, (1994) (in these works there is additional bibliography). As a rule, in linear inverse problems, certain information on the structure of the right side is supposed. Usually, in this case, additional boundary-value conditions are given. The known papers devoted to solvability of linear inverse problems are united by the method used for their solution. In this method we pass to an integral or operator equation in manifolds, where the additional boundary-value condition is given. Then we investigate this integral or operator equation.

In our monograph, first, we state some results on solvability of linear inverse problems based on this method. Namely we shall pass from the initial equation to the equation of mixed type. This method is not new; however, new results on solvability of boundary-value problems for equations of mixed type stated in Chapter 1 will allow us to obtain new results on solvability of inverse problems. Note that, actually, linear inverse problems are solved implicitly when we have solved noncoercive elliptic equations (Dubinskii, 1972) and equations of composite type and of mixed-composite type (Salakhitdinov, 1974; Dzhuraev, 1979).

First, we shall state the results on solvability of linear inverse problems for the simplest elliptic equations.

Suppose D is a bounded domain from R^n with a smooth boundary Γ, and let l be a differential operator of the first order

$$l = \alpha^k(x)\frac{\partial}{\partial x_k} + \alpha(x).$$

The sets σ_1, σ_2, and σ_0 will be similar to those defined in Section 3.1, Chapter 1.

Inverse problem I: *Find functions $u(x)$ and $q(x)$ from the relation*

$$Lu \equiv \Delta u + b(x)u = a(x)q(x) + f(x), \tag{2.1.1}$$

where $a(x)$, $b(x)$, and $f(x)$ are the functions given by the condition

$$\text{the function } q(x) \text{ belongs to the kernel of the operator } l; \tag{2.1.2}$$

and the boundary conditions

$$l_0 u\Big|_{\sigma_1\cup\sigma_2} = f_1(x), \qquad u\Big|_{\sigma_1\cup\sigma_0} = f_2(x). \tag{2.1.3}$$

Inverse problem II: *Find functions $u(x)$ and $q(x)$ connected by equation (2.1.1), condition (2.1.2), and the boundary conditions*

$$u\Big|_\Gamma = f_1(x), \qquad \frac{\partial u}{\partial \nu}\Big|_{\sigma_1} = f_2(x). \tag{2.1.4}$$

Further, when we study the inverse problems I and II, we shall assume that the functions $f_1(x)$ and $f_2(x)$ admit extension inside the domain D from the class $W_2^3(D)$. In this case, by the change $v = u - F$, where F is the desired extension, we reduce the inverse problems I and II to problems with trivial boundary conditions. Such inverse problems I and II with trivial boundary conditions we shall study in this section.

In these inverse problems I and II, condition (2.1.2) is the condition that $q(x)$ belongs to a certain class. In many works devoted to linear inverse problems we suppose that the function $q(x)$ has *a priori* a special form. As a rule, such supposition denotes that the function $q(x)$ belongs to the kernel of a certain differential operator. Further, boundary-value conditions (2.1.3) and (2.1.4) represent a combination of natural conditions for elliptic equations and the additional condition

$$u\Big|_{\sigma_1} = 0$$

in the case of the inverse problem I, or the additional condition

$$\frac{\partial u}{\partial \nu}\Big|_{\sigma_1} = 0$$

in the case of the inverse problem II. The additional condition itself in both cases is defined actually by condition (2.1.2), since the set σ_1 is determined by the coefficients of the operator l. It is possible, in particular, that the set σ_1 will be empty, and then the solution of the inverse problems I and II will be defined only by the information that the function $q(x)$ belongs to a certain class of usual elliptic boundary conditions.

We shall state now the concrete results on solvability of the inverse problems I and II.

Introduce the notations

$$\beta(x) = a(x) - \frac{l_0 a(x)}{a(x)},$$

$$l_a = l_0 + \beta(x).$$

By condition (2.1.2), the function $a(x) \cdot q(x)$ belongs to the kernel of the operator l_a. If we apply the operator l_a to equation (2.1.1), we obtain the equation

$$\tilde{L}u \equiv l_a\Big(\Delta u + b(x)u\Big) = g(x) \qquad \Big(g = l_a f\Big). \tag{2.1.5}$$

In this case, for the function $u(x)$, conditions (2.1.3) or (2.1.4) will hold. The boundary conditions are the boundary conditions investigated in Chapter 1 of the boundary-value problems I and II respectively. Thus, instead of the inverse problems I and II, we may investigate the boundary-value problems I and II for equation (2.1.5) with one unknown function $u(x)$. After determination of $u(x)$, we obtain then the function $q(x)$ and, thus, the solution of the inverse problems I and II.

Existence and uniqueness of solutions of the boundary-value problems I and II for equation (2.1.5) is provided by the results of Chapter 1. Namely, these results will be the basis of the results of existence and uniqueness of solutions of the inverse problems I and II.

For convenience, we shall repeat some results and notations of Chapter 1 (or make them more precise taking account of the concrete form of the operator L).

We introduce the notations

$$c^{ij}(x) = \beta(x)\delta^i_j - 2\alpha^j_{x_i}(x),$$

where δ^i_j is the Kronecker delta notation,

$$c^i(x) = -\Delta\alpha^i(x),$$

$$c_0(x) = \beta(x)b(x) + \alpha^k(x)b_{x_k}(x).$$

Theorem 2.1.1. *Suppose, for $x \in \bar{D}$, $\xi \in R^n$, the following conditions hold:*

$$c^{ij} = c^{ji},$$

$$\left[-\frac{1}{2}\left(\alpha^k c^{ij}\right)_{x_k} + \left(\alpha^i c^{kj}\right)_{x_k} + c^i\alpha^j\right]\xi_i\xi_j - b\left(\alpha^i\xi_i\right)^2 \geq 0,$$

$$\left(\alpha^k c_0\right)_{x_k} \geq 0;$$

for $x \in \sigma_2$, $\xi \in R^n$, the following conditions hold

$$c_0 \leq 0, \qquad c^{ij}\xi_i\xi_j \geq 0.$$

Moreover, suppose the function $a(x)$ vanishes in D only in the set of a zero measure. Then the regular solution of the inverse problem I is unique.

Proof. The regular solution of the inverse problem I will be a generalized solution of the boundary-value problem I for equation (2.1.5) (in the sense of definition 1.3.1). Substitute the function $w = l_0 u$ into the identity which defines the generalized solution (this is possible) and use the hypotheses of the theorem. Taking into account the hypotheses of the theorem and repeating the proof of Theorem 1.3.1, we may obtain that, if $f \equiv 0$, the function $u(x)$ will vanish in D almost everywhere. In this case, the function $q(x)$ will vanish in D almost everywhere also. The theorem is proved.

The theorem of existence of solution of the inverse problem I is based on Theorem 1.3.3 of existence of regular solutions of the boundary-value problem I.

As in Section 3.3 of Chapter 1, we set

$$M_0 = \max_{i,j} \max_{\bar{D}} \left| c^{ij}(x) \right|$$

and recall that the constant κ_0 is the constant from the second basic inequality for elliptic operators (see Section 3.3, Chapter 1): the constant R_1 is the constant from inequality (1.3.9) of Lemma 1.3.1.

Theorem 2.1.2. *For a certain natural k, $1 \leq k \leq n$, suppose we have the condition $|\alpha^k(x)| \geq \delta > 0$ for $x \in \bar{D}$ and suppose the intersection of the sets $\bar{\sigma}_1$ and $\overline{\sigma_0 \cup \sigma_2}$ satisfies the cone condition. Suppose, besides, for the set σ_1 the condition $\alpha^k v_k \geq \bar{\alpha}_0 > 0$ holds; for the set σ_0 the condition (1.3.7) holds; for the constants κ_0, M_0 and R_1 we have the inequality $\kappa_0 - nM_0^2 R_1^2 > 0$; and, finally, the function $a(x)$ does not vanish in $\bar{D}$, and the conditions of Theorem 2.1.1 hold. If $f(x)$ and $l_0 f(x)$ belong to the space $L_2(D)$, then there exists a solution $(u(x), q(x))$ of the inverse problem I such that $u \in W_2^2(D)$, $q \in L_2(D)$.*

Proof. As all conditions of Theorem 1.3.3 hold, then the boundary-value problem I for equation (2.1.5) has a solution $u(x)$ which attains boundary conditions (2.1.3) such that $u \in W_2^2(D)$, $l_0 u \in W_2^2(D)$. For the function $u(x)$ we have the integral identity

$$\int\limits_D \Big[\Delta u + b(x)u - f(x)\Big] l_a^* w dx = 0, \tag{2.1.6}$$

where the function $w(x)$ belongs to the set W, and l_a^* is formally conjugate to the operator l_a. Evidently, identity (2.1.6) holds and for the functions $w(x)$ from the space $\overset{0}{W}{}_2^1(D)$. We define the action of the operator l_a on functions φ from the space $L_2(D)$ as follows

$$\langle l_a \varphi, w\rangle = \int\limits_D \varphi l_a^* w dx, \tag{2.1.7}$$

where the function w belongs to the space $\overset{0}{W}{}_2^1(D)$. We set

$$\Delta u + b(x)u - f = q_1.$$

Then the function q_1 will belong to the kernel of l_a which follows from identity (2.1.6). As the function $a(x)$ does not vanish in the set $\bar{D}$, then, by the function $q_1(x)$, we may determine $q(x)$ uniquely

$$q(x) = \frac{q_1(x)}{a(x)}.$$

The function $q(x)$ will belong to the kernel of l if we define it by an equality analogous to (2.1.7).

The functions $u(x)$ and $q(x)$ will be the desired solution of the inverse problem I for equation (2.1.5). The theorem is proved.

Now, we shall state the results concerning existence and uniqueness of solutions of the inverse problem II.

We define the quadratic form $B(x,\xi)$ and the function $C(x)$ by the equalities

$$B(x,\xi) = \left[\beta(x) - \frac{1}{2}\alpha_{x_k}^k(x)\right]\xi_i^2 - \alpha_{x_i}^k(x)\xi_k\xi_i,$$

$$C(x) = \frac{1}{2}\Big(\alpha^k(x)b(x)\Big)_{x_k} - \alpha^k(x)b_{x_k}(x) - \frac{1}{2}\Delta\beta(x) - \beta(x)b(x) + \frac{1}{2}\Delta\alpha_{x_k}^k(x).$$

Theorem 2.1.3. *Suppose for $x \in \bar{D}$, $\xi \in R^n$ the following conditions*

$$B(x,\xi) \geq \beta_1 |\xi|^2, \qquad \beta_1 \geq 0,$$

$$C(x) \geq \bar{c}_0 \geq 0$$

hold, where the constants β_1 and $\bar{c}_0$ satisfy the inequality $\beta_1^2 + \bar{c}_0^2 > 0$. If a function $a(x)$ vanishes in D only in the set of zero measure, then the regular solution of the inverse problem II is unique.

Proof. The proof of this theorem is based on Theorem 1.3.5 on uniqueness of regular solutions of the boundary-value problem II.

For equations (2.1.5) in the case of the general operator l and the domain D of general geometry, we have proved in Chapter 1 existence of generalized solutions of the boundary-value problem II. As solvability of the inverse problem II is connected with solvability of the boundary-value problem II for equation (2.1.5), it is natural to suppose that the inverse problem II will have a generalized solution (the sense of generalized solution of the inverse problem II will be defined further). We shall show now that it really is so.

Suppose for all $x \in \bar{D}$ and $\xi \in R^n$ the following conditions hold

$$B(x,\xi) \geq \beta_1 |\xi|^2, \qquad \beta_1 \geq 0, \tag{2.1.8}$$

$$C(x) \geq \bar{c}_0 > 0,$$

and let the function $l_a f$ belong to the space $L_2(D)$. Then, from Theorem 1.3.7, it follows that boundary-value problem (2.1.4), (2.1.5) will have a generalized solution $u(x)$ which belongs to the space H_0. This solution will attain boundary conditions (2.1.4). For the generalized solution, the following identity will hold

$$a^0(u,w) \equiv -\int_D u_{x_i}\left(l_a^* w\right)_{x_i} dx + \int_D bu \cdot l_a^* w dx$$

$$-\int_{\sigma_2} u_{x_i} l_a^* w \nu_i d\gamma = \int_D g w dx \tag{2.1.9}$$

for each function w of the space $W_2^2(D)$.

We define the space V as the set of functions from the space W_2^1 equal to zero in the set σ_0 (the norm of an element from the space V we may define by means of the norm in the space $W_2^1(D)$). For functions $u(x)$ from the space H_0 and functions v from the space V, we define the form $\tilde{a}^0(u,v)$

$$\tilde{a}^0(u,v) \equiv -\int\limits_D u_{x_i} v_{x_i}\,dx + \int\limits_D buv\,dx - \int\limits_{\sigma_2} u_{x_i} v \nu_i\,d\gamma.$$

We suppose that the following condition holds

$$-\alpha^k \nu_k \geq \bar{\alpha}_0 > 0 \quad \text{for all} \quad x \in \sigma_2. \tag{2.1.10}$$

Then the identity

$$\left|\tilde{a}^0(u,v)\right| \leq K \left|\left|u\right|\right|_{H_0} \left|\left|v\right|\right|_V$$

holds. Therefore, the form $\tilde{a}^0(u,v)$ generates the operator A acting from the space H_0 into the space V^*. This operator is defined as follows

$$\langle Au, v\rangle = \tilde{a}^0(u,v).$$

(Here $\langle , \rangle$ is the action of the functional Au on the element v.) Note that the operator l_a^* transforms functions from the space $W_{2,0}^2(D)$ into the elements of the space V.

Now, let the function $f(x)$ be an element from the space $L_2(D)$. Then the right side of identity (2.1.9) may be written as follows

$$\int\limits_D f l_a^* w\,dx.$$

Evidently, the integral

$$\int\limits_D fv\,dx$$

may be considered as a linear continuous functional over the space V: we define this functional as $\tilde{f}$.

Identity (2.1.9) denotes that

$$\langle Au - \tilde{f}, l_a^* w \rangle = 0$$

for all functions w from the space $\overset{\circ}{W}{}^2_{2,0}(D)$. By the Dubinskii theorem on solvability of equations with operators which form a coercive pair (Dubinskii, 1972), there exists a functional q_1 which belongs to the set ker $l_a \cap V^*$ such that

$$Au = \tilde{f} + q_1.$$

The functional q_1 determines the functional q uniquely by the formula

$$\langle q, v \rangle = \left\langle q_1, \frac{v}{a} \right\rangle$$

(we suppose here that the function a is not equal to zero). Evidently, the functional q will belong to the set ker $l \cap V^*$.

These considerations allow us to formulate the following theorem.

Theorem 2.1.4. *Let conditions (2.1.8) and (2.1.10) hold and suppose that a function $a(x)$ is not equal to zero in D. Suppose the functions f and $l_a f$ belong to the space $L_2(D)$. Then there exists a function $u(x)$ from the space H_0 such that u_{x_i}, $i = 1, \dots, n$, vanish in the set σ_1 and a functional q from the set ker $l \cap V^*$ such that*

$$\tilde{a}^0(u, v) = \langle aq, v \rangle + \int\limits_D f v dx, \tag{2.1.11}$$

for each function v from the space V.

Theorem 2.1.4 and, in particular, identity (2.1.11) determine the notion of the generalized solution of the inverse problem II.

Existence of "good" solutions of the inverse problem II, as in the case of the inverse problem I, is connected with existence of regular solutions of the boundary-value problem II. However, it is evident then, if the conditions of Theorem 1.3.8 hold, the function $q(x)$ is defined only by the condition (2.1.2) (see Lemma 1.3.1). Thus, Theorem 1.3.8 does not give a satisfactory result on existence of solutions of the inverse problem II (though, in some cases, for example, for certain integro-differential equations, or in the case of more general conditions for the function $q(x)$ than condition (2.1.2), Theorem 1.3.8 gives satisfactory results on solvability of the inverse problem II). In its turn, Theorem 1.3.9 in application to the inverse problem II leads to a new result.

For convenience of comparison of the result below with Theorem 1.3.9, we shall use similar notations. Namely, let D be a domain from R^n with a smooth boundary, Q be a cylinder $D \times (0,T)$ of a finite height T, and suppose l is the following operator

$$l = K(t)\frac{\partial}{\partial t} + \alpha(t).$$

Reformulating the inverse problem II in application to the domain D and taking account of Theorem 1.3.9, we may easily prove the following theorem.

Theorem 2.1.5. *For $x \in \bar{D}$, $t \in [0,T]$, let the following conditions hold*

$$\beta(x,t) - \frac{p}{2}K_t(t) \geq \bar{\beta}_0 > 0, \qquad p = 1,3,$$

$$\beta(x,t)b(x,t) - \frac{1}{2}\Big(K(t)b(x,t)\Big)_t + K(t)b_t(x,t)$$

$$+\frac{1}{2}\beta_{tt}(x,t) - \frac{1}{2}K_{ttt}(t) \leq -\bar{\beta}_1 < 0,$$

$$b(x,t) \leq 0.$$

Suppose the function $K(t)$ is nonnegative for $t \in [0,T]$ and positive for $t = 0$, $t = T$. If the function $a(x,t)$ does not vanish in $\bar{Q}$ and the functions $f(x,t)$, $l_a f(x,t)$, $(l_a f(x,t))'_t$ belong to the space $L_2(Q)$, then there exists a solution $(u(x,t),\, q(x,t))$ of the problem II such that $u \in W_2^2(Q)$, $q \in L_2(Q)$.

Existence and uniqueness of solutions of inverse problems in the setting which corresponds to the setting of the inverse problems I and II may be established by the methods stated above. Namely, by means of transfer equations of the mixed type, and for some equations of the second order. In particular, the operator L which determines the left side of equation (2.1.1) may be a more general operator of elliptic type, operator of parabolic type, elliptic-parabolic operator (generally speaking, even an operator of the first order), an integro-differential operator of the form

$$u_t - Au - \int_0^t K(x,t,\tau)Bu(x,\tau)d\tau,$$

$$u - Au - \int_0^t K(x,t,\tau)Bu(x,\tau)d\tau$$

with a positive kernel K, an operator A elliptic with respect to spatial variables, or an elliptic-parabolic operator B with respect to spatial variables. Note that equations of the mixed type with any of these operators were not studied in Chapter 1, but in some cases they may be studied by the methods applied in Chapter 1. Parallel with the inverse problems with local conditions of overdetermination, we may investigate some inverse problems with nonlocal conditions of overdetermination if we base the methods on results of solvability of nonlocal boundary-value problems for the mixed type equations (results of Section 3.9, Chapter I). Further, in the setting of the inverse problems I and II, condition (2.1.2) may be replaced by the inhomogeneous condition

$$lq = \varphi(x)$$

with a given function $\varphi(x)$. Besides, in addition to condition (2.1.2), we may require that the function $q(x)$ partially satisfies the boundary condition necessary for its unique determination. Namely, if the set $\tilde{\sigma}_1$ is a rigorous subset of the set σ_1, we require that, in addition to condition (2.1.2), the function $q(x)$ satisfy the condition

$$q\Big|_{\tilde{\sigma}_1} = 0. \qquad (2.1.2')$$

Such condition narrows the class of admissible functions $q(x)$; however, it does not allow us to determine this function immediately. Taking into account conditions (2.1.2) and (2.1.2′) we may reduce the inverse problems I and II to the boundary-value problems for the mixed type equations with discontinuous boundary conditions. Such problems have generally not been investigate yet, but in certain simple cases they may be investigated by the methods of Chapter 1. Finally, we note that the theory of mixed type operators may be applied for research of the other linear inverse problems; for example, the problem of determination of two independent right sides.

2.1.2. On solvability of nonlinear inverse problems

We shall call the problems nonlinear if either the operator L from the left side of equation (2.1.1) is nonlinear, or the function a depends on the solution or on its derivatives, or both these conditions hold. Inverse problems with a nonlinear operator L and a linear right side may be reduced to nonlinear equations of the mixed type. We have not studied such equations in this book. However, in some cases, they may be investigated by the methods of Chapter 1. In turn, inverse problems with the function a dependent even on solution are rather complicated both in the case of the setting corresponding to the setting of the inverse problems I and II, and even in the simplest setting when we suppose that the unknown coefficient q depends only on selected variables. Nowadays, nonlinear inverse problems for elliptic equations are not investigated practically. In some works (see Beznoshchenko, 1975, 1979; Anikonov and Bubnov, 1988; Anikonov and Belov, 1989; Prilepko and Vasin, 1989; Belov and Savateev, 1991; Riganti and Savateev, 1991; Akhmatov and Belov, 1991; Prilepko and Kostin, 1992, 1993; Soloviev, 1992; Vishnevskii, 1992; Savateev, 1993; Belov, 1993; Anikonov and Vishnevskii, 1994) and some others the different nonlinear inverse problems for parabolic equations were investigated. There are a lot of works devoted to nonlinear problems for hyperbolic equations. As we shall not consider these problems, note only some monographs (Romanov, 1984; Kabanikhin, 1988; Bubnov, 1986; Savateev, 1993) where inverse problems for hyperbolic equations were considered. In all these works it was supposed that the unknown coefficient was dependent only on selected variables. The results and the methods of theory of the mixed type equations will allow us to investigate solvability of some new nonlinear inverse problems for elliptic and parabolic equations. These problems in their setting will correspond to the simplest setting of the inverse problems I and II. The problem of investigation of nonlinear inverse problems in a general setting corresponding to the setting of the inverse problems I and II at present seems to be a rather complicated problem. However, we hope that it will be solved eventually.

We consider now some concrete examples. Suppose Q is a cylinder $D \times (0,T)$, where $0 < T < +\infty$, D is a bounded domain from the space R^n with a smooth boundary Γ. In this cylinder Q we consider the equation

$$Lu \equiv u_t - \Delta u + \lambda u + q(x,t)c(x,t)u = f(x,t). \tag{2.1.12}$$

Here λ is a known positive constant, the functions $c(x,t)$ and $f(x,t)$ are known, the functions $u(x,t)$ and $q(x,t)$ are unknown. Suppose the function $q(x,t)$ satisfies the equation

$$q_t + \alpha(x,t)q = 0 \tag{2.1.13}$$

where $\alpha(x,t)$ is given and, for the function $u(x,t)$, let the boundary conditions

$$u(x,0) = u_0(x), \quad u(x,T) = u_1(x), \quad \text{for} \quad x \in D, \tag{2.1.14}$$

$$u(x,t) = \mu(x,t) \quad \text{for} \quad x \in \Gamma, \quad t \in (0,T) \tag{2.1.15}$$

hold. The functions u_0, $u_1(x)$, and $\mu(x,t)$ are known.

Now, we define more exactly the smoothness conditions for the input data for problem (2.1.12)–(2.1.15); the other conditions we shall correct further. First of all, we suppose that there exists a function $U(x,t)$ from the space $C^3(\bar{Q})$ which coincides for $x \in D$, $t = 0$ and for $x \in D$, $t = T$ with the functions $u_0(x)$ and $u_1(x)$ respectively, and for $x \in \Gamma$, $t \in (0,T)$ it coincides with the function $\mu(x,t)$ (this condition yields that the functions u_0, $u_1(x)$, and $\mu(x,t)$ are smooth functions, and the concordance conditions hold). Further, we suppose that the functions $\alpha(x,t)$, $c(x,t)$, and $c_t(x,t)$ belong to the space $C(\bar{Q})$; the function $f(x,t)$ is bounded and measurable in $\bar{Q}$.

We define the set of functions V

$$V = \Big\{ v(x,t) : v \in L_2\Big(0,T; W_2^2(D)\Big) \bigcap L_\infty(Q),$$

$$v_t \in L_2\Big(0,T; W_2^2(D)\Big) \bigcap L_\infty(Q), \quad v_{tt} \in L_2(Q) \Big\}.$$

Evidently, the set V is a Banach space. The norm in V we introduce by the equality

$$\left\|v\right\|_V = \operatorname*{vraimax}_{\bar{Q}} |v| + \operatorname*{vraimax}_{\bar{Q}} |v_t| + \left(\int\limits_Q \left[v_{tt}^2 + \left(\Delta v_t\right)^2 + \left(\Delta v\right)^2 \right] dx dt \right)^{1/2}.$$

Suppose V_0 is a subset of the set V consisting of the functions $v(x,t)$ which coincide with the function $U(x,t)$ in the boundary ∂Q of the cylinder Q. V_1 is a subset of the set V consisting of the function $v(x,t)$ which coincide with the functions $U(x,t)$ for $(x,t) \in \Gamma \times (0,T)$ and in the points (x,T) of the upper base of the cylinder Q.

We suppose that the following conditions hold

$$c(x,T) \geq c_0 > 0, \quad u_1(x) \geq k_1 > 0 \quad \text{for} \quad x \in \bar{D}, \tag{2.1.16}$$

(c_0 and k_1 are constants). Define the functions $\psi(x,t)$, $h_1(x)$, $g_1(x)$, and $h_0(x)$ as follows

$$\psi(x,t) = \exp\left(\int\limits_t^T \alpha(x,\tau) d\tau \right)$$

$$h_1(x) = \frac{f(x,T) + \Delta u_1(x) - \lambda u_1(x)}{c(x,T) u_1(x)}, \quad g_1(x) = \frac{1}{c(x,T) u_1(x)},$$

$$h_0(x) = f(x,0) + \Delta u_0(x) - \lambda u_0(x) - c(x,0)\psi(x,0)h_1(x)u_0(x).$$

We suppose that

$$q(x,t) = \psi(x,t)\Big[h_1(x) + g_1(x)u_t(x,T)\Big].$$

Consider the boundary-value problem: *find the solution of the equation*

$$L_1 u \equiv u_{tt} - \Delta u_t + \lambda u_t + cqu_t + (c_t - \alpha c)qu = f_t, \tag{2.1.12'}$$

satisfying the conditions

$$u(x,T) = u_1(x), \quad u_t(x,0) = h_0(x) - c(x,0)\psi(x,0)g_1(x)u_0(x)u_t(x,T) \tag{2.1.14'}$$

and condition (2.1.15).

Investigation of the solvability of the boundary-value problem (2.1.12′), (2.1.14′), (2.1.15) we shall carry out by combining the fixed point method and the cutoff functions method.

For a positive number $k_0 \geq k_1$, we define the function $G(\xi)$:

$$G(\xi) = \begin{cases} 0 \ \text{if}\, \xi < 0, \\ \xi \ \text{if}\, \xi \in [0, k_0], \\ k_0 \ \text{if}\, \xi > k_0. \end{cases}$$

Further, for the functions $v(x,t)$ from the set V we define the functions $q_v(x,t)$ and $F_v(x,t)$:

$$q_v(x,t) = \psi(x,t)\Big[h_1(x) - g_1(x)\big|v_t(x,T)\big|\Big],$$

$$F_v(x,t) = f_t(x,t) - \Big[c_t(x,t) - \alpha(x,t)c(x,t)\Big]q_v(x,t)G\Big(v(x,t)\Big).$$

We consider the following boundary-value problem: *find a solution of the equation*

$$L_0 u \equiv u_{tt} - \Delta_t + \lambda u_t + c q_v u_t = F_v, \tag{2.1.12''}$$

satisfying conditions (2.1.14′) and (2.1.15).

Suppose the following conditions hold:

$$c(x,t) \geq 0 \quad \text{for} \quad (x,t) \in \bar{Q}, \tag{2.1.17}$$

$$h_0(x) \le 0, \quad 0 \le c(x,0)\psi(x,0)u_0(x) < c(x,T)u_1(x) \text{ for } x \in \bar{D}, \tag{2.1.18}$$

$$\mu_t(x,0) = h_0(x) - c(x,0)\psi(x,0)g_1(x)u_0(x)\mu_t(x,T) \text{ for } x \in \Gamma. \tag{2.1.19}$$

As the function $v(x,t)$ belongs to the set V, then the function $F_v(x,t)$ will be a bounded and measurable set in $\bar{Q}$. Taking this fact into account and using conditions (2.1.17)–(2.1.19), we may show the existence of the solution $u(x,t)$ of problem (2.1.12″), (2.1.14′), (2.1.15) such that the estimates

$$\operatorname{vraimax}|u| \le K_0, \qquad \operatorname{vraimax}|u_t| \le K_1, \tag{2.1.20}$$

$$\int \left[u_{tt}^2 + \left(\Delta u_t\right) + \left(\Delta u\right)\right] dxdt \le R \tag{2.1.21}$$

will hold. Here K_0, K_1 and R are certain constants dependent on the function $v(x,t)$ and on input data of the problem. Really, equation (2.1.12″) is a linear parabolic equation for the function $u_t(x,t)$. Boundary conditions and the conditions for the input data mentioned above allow us to find this function. After determining the function $u_t(x,t)$, we find then the function $u(x,t)$. Further, the second of estimates (2.1.20) is the corollary of the maximum principle estimate for parabolic equations. The first estimate follows from the second one. Estimate (2.1.21) is almost the standard integral estimate for solutions of parabolic equations (further, we shall need the exact values of the conditions K_1, K_2 and R_0; therefore, we shall produce a rigorous proof of estimates (2.1.20) and (2.1.21)).

So, for each function $v(x,t)$ from the set V, we define the function $u(x,t)$ which solves problem (2.1.12″), (2.1.14′), (2.1.15) and belongs to the set V_1. Thus, we have defined the operator A which maps the function $v(x,t)$ into the function $u(x,t)$. This operator acts from the set V into itself and from the set V_1 into itself also. We show now that the operator A has fixed points in the set V (note that if the function $u(x,t)$ is a fixed point of the operator A, then this function will belong to the set V_1).

In order that the operator A has fixed points in the set V it suffices to show that it is completely continuous in the set V and it maps a certain closed convex bounded subset of the set V into itself.

First, we show that the operator A is completely continuous in V.

Suppose a sequence of functions $\{v_m(x,t)\}_{m=1}^{\infty}$ is bounded in V. Then there exists a subsequence $\{v_{m_k}(x,t)\}_{k=1}^{\infty}$ and a function $v(x,t)$ from the set V such that

$$v_{m_k}(x,t) \underset{k\to\infty}{\to} v(x,t) \quad \text{almost everywhere in} \quad Q,$$

$$v_{m_k t}(x,t) \underset{k\to\infty}{\to} v_t(x,t) \quad \text{almost everywhere in} \quad Q,$$

$$v_{m_k t}(x,T) \underset{k\to\infty}{\to} v_t(x,T) \quad \text{almost everywhere in} \quad D.$$

Suppose the functions $u_{m_k}(x,t)$ are the functions from $v_{m_k}(x,t)$ generated by means of the operator A. Then, for the functions $u_{m_k}(x,t)$, the estimates (2.1.20) and (2.1.21) will hold uniform with respect to k. From these estimates it follows that, from the sequence $\{u_{m_k}(x,t)\}_{k=1}^{\infty}$, we may choose a new subsequence (we shall save the notation of this sequence) such that, for a certain function $u(x,t)$ from the set V, the following convergences will hold

$$u_{m_k tt} \underset{k\to\infty}{\to} u_{tt} \quad \text{weakly in} \quad L_2(Q),$$

$$\Delta u_{m_k t} \underset{k\to\infty}{\to} u_t \quad \text{weakly in} \quad L_2(Q),$$

$$u_{m_k}(x,t) \underset{k\to\infty}{\to} u(x,t) \quad \text{almost everywhere in} \quad Q,$$

$$u_{m_k t}(x,t) \underset{k\to\infty}{\to} u_t(x,t) \quad \text{almost everywhere in} \quad Q,$$

$$u_{m_k t}(x,T) \underset{k\to\infty}{\to} u_t(x,T) \quad \text{almost everywhere in} \quad Q.$$

Evidently, the limit functions $u(x,t)$ and $v(x,t)$ will be connected by equation (2.1.12″). For the function $u(x,t)$, boundary conditions (2.1.14′) and (2.1.15) will hold true. Then, the difference $u_{m_k} - u$ will satisfy an equation of the type (2.1.12″) and homogeneous boundary conditions (2.1.14′) and (2.1.15).

Applying the maximum principle estimate for parabolic equations and taking into account the convergences for the functions $v_{m_k}(x,t)$ pointed out above, we see that

$$\operatorname*{vraimax}_{\bar{Q}} \left| u_{m_k t} - u_t \right|_{k \to \infty} 0.$$

From this convergence of derivatives follows the convergence for the functions

$$\operatorname*{vraimax}_{\bar{Q}} \left| u_{m_k} - u \right|_{k \to \infty} 0.$$

Finally, the integral convergence

$$\int\limits_Q \left[\left(u_{m_k tt} - u \right)^2 + \left(\Delta u_{m_k t} - \Delta u_t \right)^2 + \left(\Delta u_{m_k} - \Delta u \right)^2 \right] dxdt_{k \to \infty} 0$$

is a corollary of estimate (2.1.21) and the convergences of the functions $u_{m_k}(x,t)$ and $v_{m_k}(x,t)$ already pointed out.

From these convergences it follows that for each bounded sequence of functions $\{v_m(x,t)\}_{m=1}^{\infty}$ which belong to V, from the sequence $\{Av_m\}_{m=1}^{\infty}$, we may select a strongly converging subsequence. It denotes that the operator A is completely continuous in the space V.

Now, we shall prove that the operator A maps a certain closed convex subset W of the set V into itself.

Suppose M_0, M_1, and R_0 are certain positive numbers and W is the set defined as follows

$$W = \Bigg\{ v \in V_1 : \operatorname*{vraimax}_{\bar{Q}} |v| \le M_0, \quad \operatorname*{vraimax}_{\bar{Q}} |v_t| \le M_1,$$

$$\int\limits_Q \left[v_{tt}^2 + + \left(\Delta v \right)^2 \right] dxdt \le R_0 \Bigg\}.$$

We show now that the numbers M_0, M_1, and R_0 may be chosen so that the set W will be produced.

We define the constants which we shall need. We set

$$C_1 = \operatorname*{vraimax}_{\bar{Q}} \left| f_t(x,t) \right|,$$

$$C_2 = \operatorname*{vraimax}_{\bar{Q}} \left[\left| c_t(x,t) - \alpha(x,t)c(x,t) \right| \cdot \left| h_1(x) \right| \cdot \psi(x,t) \right],$$

$$C_3 = \max_{\bar{Q}} \frac{\left| c_t(x,t) - \alpha(x,t)c(x,t) \right| \cdot \psi(x,t)}{c(x,T)u_1(x)},$$

$$\bar{h}_0 = \operatorname*{vraimax}_{\bar{D}} \left| h_0(x) \right|, \qquad \mu_1 = \max_{x \in \Gamma, t \in [0,T]} \left| \mu_t(x,t) \right|,$$

$$\lambda_0 = \max_{\bar{D}} \left| c(x,0)\psi(x,0)g_1(x)u_0(x) \right|.$$

Suppose that the function $v(x,t)$ belongs to the set W. The classical estimate of the maximum principle for parabolic equations gives the following inequality for the solution of problem (2.1.12″), (2.1.14′), (2.1.15):

$$\operatorname*{vraimax}_{\bar{Q}} |u_t| \le \max \left(\frac{C_1 + C_2 k_0 + C_3 M_1 k_0}{\lambda},\ \bar{h}_0 + \lambda_0 \operatorname*{vraimax}_{\bar{Q}} |u_t|,\ \mu_1 \right).$$

From this inequality follows the next inequality

$$\operatorname*{vraimax}_{\bar{Q}} |u_t| \le \max \left(\frac{C_1 + C_2 k_0}{\lambda},\ \bar{h}_0,\ \mu_1 \right) + \frac{C_3 M_1 k_0}{\lambda} + \lambda_0 \operatorname*{vraimax}_{\bar{Q}} |u_t|.$$

Condition (2.1.18) denotes that $\lambda_0 < 1$. Therefore, $\operatorname*{vraimax}_{\bar{Q}} |u_t|$ may be estimated as follows:

$$\operatorname*{vraimax}_{\bar{Q}} |u_t| \le \frac{1}{1 - \lambda_0} \max \left(\frac{C_1 + C_2 k_0}{\lambda},\ \bar{h}_0,\ \mu_1 \right) + \frac{C_3 M_1 k_0}{\lambda(1 - \lambda_0)}. \qquad (2.1.22)$$

This inequality gives estimate (2.1.20).

Denote

$$N_0 = \frac{1}{1-\lambda_0} \max\left(\frac{C_1 + C_2 k_0}{\lambda},\ \bar{h}_0,\ \mu_1\right), \quad N_1 = \frac{C_3 k_0}{\lambda(1-\lambda_0)}.$$

We suppose that

$$N_1 < 1 \tag{2.1.23}$$

and the number M_1 which defines the set W is such that

$$M_1 \geq \frac{N_0}{1 - N_1}.$$

Then, inequality (2.1.22) yields inequality

$$\underset{\bar{Q}}{\operatorname{vraimax}} |u_t| \leq M_1. \tag{2.1.24}$$

Note, further, that estimate (2.1.24) allows us to estimate $\underset{\bar{Q}}{\operatorname{vraimax}} |u|$:

$$\underset{\bar{Q}}{\operatorname{vraimax}} |u| \leq \max_{\bar{D}} u_1(x) + TM_1.$$

If we choose a constant M_0 so that

$$M_0 \geq \max_{\bar{D}} u_1(x) + TM_1,$$

then, if v belongs W, then

$$\underset{\bar{Q}}{\operatorname{vraimax}} |u| \leq M_0. \tag{2.1.25}$$

Equation (2.1.12″) may be written in the form

$$u_{tt} - \Delta u_t = \Phi. \tag{2.1.26}$$

Taking into account inequalities (2.1.24), (2.1.25), the conditions imposed on coefficients of equations (2.1.12), (2.1.13) and functions $f(x,t)$, $u_0(x)$, $u_1(x)$, and the fact that $v(x,t)$ belongs to W, we obtain the estimate for the function Φ

$$\operatorname*{vraimax}_{\bar{Q}} |\Phi| \le R_1. \tag{2.1.27}$$

Here R_1 is a certain constant dependent only on input data of the problem and on the constants M_0 and M_1 (the exact value of R_1 readers may obtain for themselves). We multiply equation (2.1.26) from the right and from the left by the function $u_t - U_t$ and integrate the result along the cylinder Q. Relatively straightforward calculations, estimates (2.1.24), (2.1.26), and the conditions imposed on the coefficients of equations (2.1.12), (2.1.13), and the functions $f(x,t)$, $u_0(x)$, $u_1(x)$ yield the estimate

$$\sum_{i=1}^{n} \int_Q u_{x_i t}^2 dx dt \le R_2, \tag{2.1.28}$$

where R_2 is a certain constant dependent on input data of the problem and on M_0 and M_1.

Estimate (2.1.28) allows us to obtain the following estimate. Namely, integrating by parts in the equality

$$\int_Q \left(u_{tt} - \Delta u_t\right)\left(u_{tt} - U_{tt}\right) dx dt = \int_Q \Phi\left(u_{tt} - U_{tt}\right) dx dt,$$

taking account of estimates (2.1.27), (2.1.28), condition (2.1.14′) and the conditions imposed on the coefficients of equations (2.1.12) and (2.1.13) and the functions $f(x,t)$, $u_0(x)$, $u_1(x)$, we obtain the estimate

$$\int_Q u_{tt}^2 dx dt \le R_3. \tag{2.1.29}$$

Here the constant R_3 is dependent only on input data of the problem and the numbers M_0 and M_1.

The following estimate

$$\int_Q \left(\Delta u_t\right)^2 dx dt \le R_4 \tag{2.1.30}$$

now is evident; the constant R_4 again depends only on input data of the problem and the numbers M_0 and M_1. Finally, estimate (2.1.30) yields the estimate

$$\int\limits_Q \left(\Delta u\right)^2 dxdt \le R_5. \tag{2.1.31}$$

In order to obtain this estimate, we need to integrate the equality

$$t\frac{\partial}{\partial t}\left(\Delta u\right)^2 = 2t\Delta u \cdot \Delta u_t$$

along the cylinder and to use estimate (2.1.30). Like all the constants R_1–R_4 above, the constant R_5 will depend only on input data and the numbers M_0 and M_1.

We suppose that the number R_0 which defines the set W is such that

$$R_0 \ge R_3 + R_4 + R_5.$$

Then, solutions of problem (2.1.12″), (2.1.14′), (2.1.15) will satisfy the estimate

$$\int\limits_Q \left[u_{tt}^2 + \left(\Delta u_t\right)^2 + \left(\Delta u\right)^2\right] dxdt \le R_0. \tag{2.1.32}$$

In other words, for solutions of problem (2.1.12″), (2.1.14′), (2.1.15) we establish estimate (2.1.21). Estimates (2.1.24), (2.1.25), and (2.1.32) denote that the operator A transfers the set W into itself (if the numbers M_0, M_1 and R_0 are chosen as above).

So, we have proved that for the operator A all the conditions of the Schauder theorem on a fixed point are fulfilled. By this theorem, there exists a function $u(x,t)$ which attains boundary conditions (2.1.14′), (2.1.15) and which solves the equation

$$u_{tt} - \Delta u_t + \lambda u_t + cq_u \cdot u_t + (c_t - \alpha c)q_u \cdot G(u) = f_t. \tag{2.1.33}$$

Suppose, in addition to the conditions above, the following conditions hold

$$f_t(x,t) \le 0, c_t(x,t) - \alpha(x,t)c(x,t) \ge 0 \text{ for } x \in \bar{Q}, \tag{2.1.34}$$

$$u_1(x) \leq u_0(x) \leq k_0, \quad h_0(x) \leq 0 \quad \text{for} \quad x \in \bar{D}, \tag{2.1.35}$$

$$\mu_1(x,t) \leq 0 \quad \text{for} \quad x \in \Gamma, \quad t \in [0,T]. \tag{2.1.36}$$

First, we note that under these conditions solutions of equation (2.1.33) satisfying conditions (2.1.14′) and (2.1.15) will satisfy the property

$$u_t(x,t) \leq 0 \quad \text{for almost all} \quad (x,t) \in \bar{Q}. \tag{2.1.37}$$

Really, from conditions (2.1.16)-(2.1.18), (2.1.34), and (2.1.36) it follows that the positive maximum (essential) of the function $u_t(x,t)$ cannot be at the points (x,t) of the cylinder Q such that $x \in \bar{D}$, $t > 0$. However, at the points $(x,0)$ of the lower base of the cylinder Q, the desired positive maximum also cannot exist. This is so, since conditions (2.1.16)-(2.1.18), (2.1.35) together with second conditions from (2.1.14′) yield the inequalities

$$0 < u_t(x,0) < u_t(x,T)$$

at such points. These inequalities denote that the positive maximum of the function $u_t(x,t)$ cannot be in the points of the lower base of Q.

As the positive maximum of the function $u_t(x,t)$ does not exist in Q, then the property (2.1.37) holds.

Nonnegativeness of the derivative $u_t(x,t)$ denotes that the solution of equation (2.1.33) solves the equation

$$u_{tt} - \Delta u_t + \lambda u_t + cqu_t + (c_t - \alpha c)qG(u) = f_t, \tag{2.1.33′}$$

and that the inequalities

$$u_1(x) \leq u(x,t) \leq u(x,0) \tag{2.1.38}$$

hold for almost all points (x,t) of the cylinder $\bar{Q}$.

Now, we integrate equation (2.1.33′) with respect to the variable t from 0 to T. Straightforward transformations lead us to the equality

$$u_t(x,0) - \Delta u(x,0) + \lambda u(x,0) + c(x,0)\psi(x,0)\Big[h_1(x) + g_1(x)u_t(x,T)\Big]u(x,0)$$

$$= f(x,0) + \int\limits_0^T (c_t - \alpha c)q\Big[G(u) - u\Big]\,dt.$$

If we subtract the second equality (2.1.14′) from this equality, we obtain the following equality

$$-\Delta\Big[u(x,0) - u_0(x)\Big] + \lambda\Big[u(x,0) - u_0(x)\Big] + cq\Big[u(x,0) - u_0(x)\Big]$$

$$= \int\limits_0^T (c_t - \alpha c)q\Big[G(u) - u\Big]\,dt. \tag{2.1.39}$$

From equalities (2.1.38) follows positiveness of the solution $u(x,t)$ of problem (2.1.33), (2.1.14′), (2.1.15). However, for positive functions $u(x,t)$, we have the inequality

$$G(u) - u \le 0.$$

Taking into account condition (2.1.34) and equality (2.1.39), we obtain the inequality

$$-\Delta\Big[u(x,0)-u_0(x)\Big]+\lambda\Big[u(x,0)-u_0(x)\Big]+cq\Big[u(x,0)-u_0(x)\Big] \le 0. \tag{2.1.40}$$

As values of the functions $u(x,0)$ and $u_0(x)$ coincide for $x \in \Gamma$, from the maximum principle for elliptic operators and from inequality (2.1.40) follows the inequality

$$u(x,0) \le u_0(x) \quad \text{for almost all} \quad x \in \bar{D}.$$

If we recall now inequalities (2.1.38), we see that solutions of equation

(2.1.33) satisfying conditions (2.1.14′) and (2.1.15) will satisfy the inequalities

$$k_1 \le u(x,t) \le k_0 \quad \text{for almost all} \quad (x,t) \in \bar{Q}. \tag{2.1.41}$$

These inequalities make estimate (2.1.25) more precise. Moreover, they denote that solutions of equation (2.1.33) satisfying conditions (2.1.14′) and (2.1.15) satisfy the equalities

$$G(u) \equiv u,$$

$$-\Delta\Big[u(x,0)-u_0(x)\Big]+\lambda\Big[u(x,0)-u_0(x)\Big]+cq\Big[u(x,0)-u_0(x)\Big]\equiv 0.$$

Again, taking the maximum principle into account, we see that solutions of problem (2.1.33), (2.1.14′), (2.1.15) satisfy the condition

$$u(x,0)=u_0(x).$$

Finally, inequalities (2.1.41) denote that equation (2.1.33′) transforms into equation (2.1.12′). Integrating this equation with respect to the variable t from the moving point to the point T, we may easily show that solutions of problem (2.1.33), (2.1.14′), (2.1.15) will solve problem (2.1.12)–(2.1.15) also.

For convenience, we shall formulate the result that we established as the following theorem.

Theorem 2.1.6. *Suppose the function $f(x,t)$ is such that $f(x,t)$ and $f_t(x,t)$ are bounded and measurable in the cylinder $\bar{Q}$; let the functions $u_0(x)$, $u_1(x)$ and $\mu(x,t)$ be such that there exists a function $U(x,t)$ from the space $C^3(\bar{Q})$, which coincides with the functions $u_0(x)$ and $u_1(x)$ for $x \in D$, $t=0$ and $x \in D$, $t=T$ respectively and with the function $\mu(x,t)$ for $x \in \Gamma$, $t \in (0,T)$. Further, suppose the functions $\alpha(x,t)$, $c(x,t)$, and $c_t(x,t)$ belong to the space $C(\bar{Q})$ and suppose for $x \in \bar{D}$, $t \in [0,T]$ the conditions*

$$c(x,t)\le 0, \quad c_t(x,t)-\alpha(x,t)c(x,t)\ge 0, \quad f_t(x,t)\le 0;$$

$$c(x,T) \geq c_0 > 0, \qquad 0 < k_1 \leq u_1(x) \leq u_0(x) \leq k_0;$$

$$h_1(x) \geq 0, \quad h_0(x) \leq 0, \quad c(x,0)\psi(x,0)u_0(x) < c(x,T)u_1(x);$$

$$k_0 \frac{\Big[c_t(x,t) - \alpha(x,t)c(x,t)\Big]\psi(x,t)}{c(x,T)u_1(x)} < \lambda\Big[1 - c(x,0)\psi(x,0)g_1(x)u_0(x)\Big]$$

hold. Finally, suppose, for $x \in \Gamma$, $t \in [0,T]$, the condition

$$\mu_t(x,t) \leq 0$$

holds; and, for $x \in \Gamma$, the condition

$$\mu_t(x,0) = h_0(x) - c(x,0)\psi(x,0)g_1(x)u_0(x)\mu_t(x,T)$$

holds.

Then, inverse problem (2.1.12)-(2.1.15) has a solution $u(x,t)$ which belongs to the set V_0.

Inverse problem (2.1.12)-(2.1.15) in its setting comes from the boundary-value problem II for the parabolic equation differentiated with respect to t. Thus, this problem is an inverse problem of the same type as the inverse problem II. By similar methods to those considered above, we investigate the inverse problem of the type of the inverse problem I for equation (2.1.12). Namely, we show how to prove solvability of the following inverse problem: *find functions $u(x,t)$ and $q(x,t)$ connected by equation (2.1.12) if $q(x,t)$ solves equation (2.1.13), and the function $u(x,t)$ satisfies condition (2.1.15) and*

$$u(x,0) = u_0(x), \quad u_t(x,0) = u_1(x) + \gamma u_t(x,T) \ \text{ for } \ x \in D. \qquad (2.1.14'')$$

Note that this inverse problem is a problem with a nonlocal condition. If $\gamma = 0$, this problem is trivial, but if $\gamma \neq 0$, this is not so.

We shall assume that for the coefficients $c(x,t)$, $\alpha(x,t)$ and the function $f(x,t)$, similar smoothness conditions hold as in conditions of Theorem 2.1.6. As before, we shall assume that there exists a function $U(x,t)$ from the space $C^3(\bar{Q})$ which coincides with the functions $u_0(x)$ and $\mu(x,t)$ for $x \in D$, $t=0$ and $x \in \Gamma$, $x \in (0,T)$ respectively. Also, we assume that $u_1(x)$ belongs to the space $C^1(\bar{D})$.

Suppose the following conditions hold:

$$c(x,0) \geq c_0 > 0, \qquad 0 < k_0 \leq u_0(x) \leq k_1, \tag{2.1.42}$$

where c_0, k_0, and k_1 are constants. We define the functions $\varphi(x,t)$, $\tilde{h}_0(x)$, and $g_0(x)$ as follows:

$$\varphi(x,t) = \exp\left(-\int_0^t \alpha(x,\tau)d\tau\right),$$

$$\tilde{h}_0(x) = \frac{f(x,0) + \Delta u_0(x) - \lambda u_0(x)}{c(x,0)u_0(x)}, \qquad g_0(x) = -\frac{1}{c(x,0)u_0(x)}.$$

We set

$$\tilde{g}(x,t) = \varphi(x,t)\left[\tilde{h}_0(x) + g_0(x)u_t(x,0)\right]$$

and consider the boundary-value problem: *find a solution of the equation*

$$u_{tt} - \Delta u_t + \lambda u_t + c\tilde{q}u_t + (c_t - \alpha c)\tilde{q} \cdot u = f_t \tag{2.1.43}$$

satisfying conditions (2.1.14″) and (2.1.15).

As before, we introduce the function $G(\xi)$, but now, by means of the number k_1; also, we define for the functions $v(x,t)$ from the set V the functions $\tilde{q}_v(x,t)$ and $F_v(x,t)$. We consider the linear problem: *find the solution of the equation*

$$u_{tt} - \Delta u_t + \lambda u_t + c\tilde{q}_v \cdot u_t = F_v, \tag{2.1.43′}$$

satisfying conditions (2.1.14″) and (2.1.15).

Suppose the following conditions

$$0 \leq \gamma < 1, \quad \tilde{h}_0(x) \geq 0, \quad u_1(x) \leq 0 \text{ for } x \in \bar{D},$$

$$\mu_t(x,0) = u_1(x) + \gamma \mu_t(x,T) \quad \text{for} \quad x \in \Gamma$$

and the condition (2.1.17) hold. As before, linear problem (2.1.43′), (2.1.14″), (2.1.15) will have a solution $u(x,t)$ from the set V, and the operator A, which transforms functions $v(x,t)$ from the set V into the function $u(x,t)$, will be completely continuous.

Suppose $\tilde{W}$ is the set of functions which belong to the set V, attain boundary conditions (2.1.14″) and (2.1.15), and such that

$$\operatorname*{vraimax}_{\bar{Q}} |v| \leq M_0, \qquad \operatorname*{vraimax}_{\bar{Q}} |v_t| \leq M_1,$$

$$\int\limits_Q \left[v_{tt}^2 + \left(\Delta v_t\right)^2 + \left(\Delta v\right)^2 \right] dx dt \leq R_0.$$

We set

$$\tilde{C}_2 = \operatorname*{vraimax}_{\bar{Q}} \left[\left| c_t(x,t) - \alpha(x,t) c(x,t) \right| \left| \tilde{h}_0(x) \right| \varphi(x,t) \right],$$

$$\tilde{C}_3 = \max_{\bar{Q}} \frac{\left| c_t(x,t) - \alpha(x,t) c(x,t) \right| \varphi(x,t)}{c(x,0) u_0(x)},$$

$$\tilde{k}_1 = \max_{\bar{D}} \left| u_1(x) \right|, \qquad \tilde{N}_0 = \frac{1}{1-\gamma} \max \left(\frac{C_1 + \tilde{C}_2 k_1}{\lambda}, \tilde{k}_1, \mu_1 \right)$$

(the constants C_1 and μ_1 are defined similarly as before),

$$\tilde{N}_1 = \frac{\tilde{C}_3 k_1}{\lambda(1-\gamma)}.$$

If the condition

$$\tilde{N}_1 < 1$$

holds and M_1 is a number such that

$$M_1 \geq \frac{\tilde{N}_0}{1 - \tilde{N}_1},$$

then, solutions of problem (2.1.43′), (2.1.14″), (2.1.15) will satisfy the estimate

$$\operatorname*{vraimax}_{\bar{Q}} |u_t| \leq M_1.$$

Moreover, as before, we may show that solutions of problem (2.1.43′), (2.1.14″), (2.1.15) will satisfy the property

$$u_t(x,t) \leq 0 \quad \text{for almost all} \quad (x,t) \in \bar{Q}.$$

Hence it follows that the function $u(x,t)$ satisfies the inequality

$$u(x,t) \geq k_0 - M_1 T.$$

Suppose the following condition holds

$$k_0 \geq \frac{\tilde{N}_0 T}{1 - \tilde{N}_1}.$$

Then we may choose a constant M_1 so that

$$u_t(x,t) \geq 0 \quad \text{for almost all} \quad (x,t) \in \bar{Q}.$$

Taking these inequalities into account, we may show that there exist such constants M_0 and R_0 that the operator A maps $\tilde{W}$ into itself. As before, we show that fixed points of the operator A are solutions of inverse problem (2.1.12), (2.1.13), (2.1.14″), (2.1.15).

The result obtained we formulate as the theorem.

Theorem 2.1.7. *Suppose a function $f(x,t)$ is such that $f(x,t)$ and $f_t(x,t)$ are bounded and measurable in the cylinder $\bar{Q}$. We suppose that functions $u_0(x)$ and $\mu(x,t)$ are such that there exists a function $U(x,t)$ from $C^3(\bar{Q})$ which coincides with $u_0(x)$ and $\mu(x,t)$ for $x \in D$, $t=0$ and $x \in \Gamma$, $t \in (0,T)$ respectively. Further, let a function $u_1(x)$ belong to $C^1(\bar{D})$ and functions $\alpha(x,t)$, $c(x,t)$ and $c_t(x,t)$ belong to $C(\bar{Q})$; and suppose, for $x \in \bar{D}$, $t \in [0,T]$, the following conditions hold:*

$$c(x,t) \geq 0, \quad c_t(x,t) - \alpha(x,t)c(x,t) \geq 0, \quad f_t(x,t) \leq 0,$$

$$c(x,0) \geq c_0 > 0, \qquad 0 < k_0 \leq u_0(x) \leq k_1,$$

$$\tilde{h}_0(x) \geq 0, \qquad u_1(x) \leq 0.$$

Suppose, for $x \in \Gamma$, $t \in [0,T]$, we have

$$\mu_t(x,t) \leq 0$$

and, for $x \in \Gamma$, we have

$$\mu_t(x,0) = u_1(x) + \gamma\mu_t(x,T).$$

Finally, we suppose that

$$0 \leq \gamma < 1, \qquad \tilde{N}_1 < 1, \qquad k_0 \geq \frac{\tilde{N}_0 T}{1-\tilde{N}_1}.$$

Then inverse problem (2.1.12), (2.1.13), (2.1.14″), (2.1.15) has the solution $\{u(x,t),\ q(x,t)\}$ such that $u(x,t) \in V$, $q(x,t) \in L_\infty(0,T;L_2(D))$; $u(x,t)$ attains the boundary conditions given.

By the methods stated above, we may investigate problems for elliptic equations. So, passing to the mixed type equation, we may investigate the following inverse problems in the cylinder Q for elliptic equations:

find functions $u(x,t)$ and $q(x,t)$ connected by the equation

$$u_{tt} + \Delta u - \lambda u + q(x,t)c(x,t)u = f(x,t), \tag{2.1.44}$$

so that the function $q(x,t)$ solves (2.1.13) and the conditions (2.1.15) and

$$u(x,0) = u_0(x), \quad u_t(x,0) = v_0(x), \quad u_t(x,T) = v_1(x) \tag{2.1.45}$$

hold;

find functions $u(x,t)$ and $q(x,t)$ connected by equation (2.1.44) if the function $q(x,t)$ solves equation (2.1.13) and conditions (2.1.15) and

$$u(x,0) = u_0(x), \quad u_t(x,0) = v_0(x), \quad u(x,T) = u_1(x) \tag{2.1.46}$$

hold.

Problem (2.1.44), (2.1.15), (2.1.45) is the inverse problem of the type of the inverse problem I. Problem (2.1.44), (2.1.15), (2.1.46) is the inverse problem of the type of the inverse problem II. Both these problems may be investigated similarly: we investigate the equation which is obtained by twice differentiating equation (2.1.44) with respect to the variable t and equation (2.1.13). The function $q(x,t)$ for both problem is as follows:

$$q(x,t) = \varphi(x,t)\Big[h_0(x) + g_0(x)u_{tt}(x,0)\Big],$$

where $h_0(x)$ and $g_0(x)$ are defined as below

$$h_0(x) = \frac{f(x,0) + \Delta u_0(x) - \lambda u_0(x)}{c(x,0)u_0(x)}, \qquad g_0 = -\frac{1}{c(x,0)u_0(x)}.$$

Naturally, we suppose that

$$c(x,0) \geq c_0 > 0, \quad u_0(x) \geq k_0 > 0 \quad \text{for} \quad x \in \bar{D}.$$

Solvability of other inverse problems, among them some that are essentially more complicated (for example, for the nonlinear equations with respect to u) may be established also by the methods stated in this book. We keep in mind the methods based on the use of the theory of composite type equations.

We have established solvability of the inverse problems in Theorems 2.1.6 and 2.1.7 in the spaces constructed on the base of the spaces L_∞ and L_2.

Also, we prove existence of solutions in the spaces constructed on the base of Hölder spaces by the methods based on using the theory of composite type equations.

Further, the technique of passing to composite type equations gives us the possibility to establish new theorems of uniqueness of inverse problem solutions. Now, we show this using the inverse problem (2.1.12)–(2.1.15).

We define the constant C_4 as follows:

$$C_4 = \max_{\bar{Q}} \left[\frac{c(x,t)\psi(x,t)}{c(x,T)u_1(x)} \right].$$

The constants C_2 and C_3 which we shall use below will be just the same as the constants that were used while proving Theorem 2.1.16.

Theorem 2.1.8. *Suppose that* $\{u_1(x,t), q_1(x,t)\}$, $\{u_2(x,t), q_2(x,t)\}$ *are two solutions of inverse problem (2.1.12)–(2.1.15) such that for the functions* $u_1(x,t)$ *and* $u_2(x,t)$ *the inequalities*

$$\operatorname*{vraimax}_{\bar{Q}} |u_i| \le \beta_0, \qquad \operatorname*{vraimax}_{\bar{Q}} |u_{it}| \le \beta_1, \quad i = 1,2,$$

hold; and the functions $u_i(x,t)$, $i = 1,2$ *lie in* V_0. *If the condition*

$$C_2 T + C_3(\beta_0 + \beta_1 T) + C_4 \beta_1 < 1,$$

holds, then the functions $u_1(x,t)$ *and* $u_2(x,t)$; $q_1(x,t)$ *and* $q_2(x,t)$ *coincide almost everywhere in* $\bar{Q}$.

Proof. The function $w(x,t) = u_1(x,t) - u_2(x,t)$ satisfies the equation

$$w_{tt} - \Delta w_t + \lambda w_t + cq(u_2)w_t = -(c_t - \alpha c)\psi h_1 w +$$

$$(c_t - \alpha c)\psi g_1 u_{2t}(x,T)w - c\psi g_1 u_{1t} w_t(x,T) - (c_t - \alpha c) g_1 u_1 w_t(x,T),$$

and the conditions

$$w(x,0) = 0, \qquad w(x,T) = 0,$$

$$w(x,T)\Big|_{\Gamma\times(0,T)} = 0.$$

Applying to the function w the estimate of the maximum principle for parabolic equations and the evident estimate

$$\operatorname*{vraimax}_{\bar{Q}} |w| \leq T \cdot \operatorname*{vraimax}_{\bar{Q}} |w_t|,$$

we obtain the inequality

$$\operatorname*{vraimax}_{\bar{Q}} \Big|w_t(x,t)\Big| \leq \Big[C_2 T + C_3(\beta_0 + \beta_1 T) + C_4\beta_1\Big] \operatorname*{vraimax}_{\bar{Q}} \Big|w_t(x,t)\Big|.$$

By the theorem condition, this inequality yields $w(x,t) \equiv 0$ almost everywhere in the cylinder $\bar{Q}$. From here follows the theorem statement. The theorem is proved.

Remark. The conditions of Theorems 2.1.6 and 2.1.7 which connect the different functions and constants through inequalities seem to be rather tedious. However, it is evident, the classes of equations which satisfy these conditions are not empty. Moreover, in some cases these conditions are sufficient, but not minimal. In some cases these estimates may be improved.

2.2. SOLVABILITY OF NONCLASSICAL BOUNDARY-VALUE PROBLEMS FOR EQUATIONS OF SECOND ORDER AND THIRD ORDER

Theory of equations of mixed type may be applied both for investigation of inverse problems for partial differential equations and for other parts of the general theory of differential equations. So, equations of mixed type may be applied as regularizators for investigation of second order equations. General setting of the boundary-value problems I and II for mixed type equations may generate new settings of boundary-value problems for nonclassical second order equations. Finally, the theory of mixed type equations may be applied for research of special boundary-value problems – those problems with an oblique derivative – for various equations of second and higher orders.

In this paragraph we shall give some examples of the use of the theory of boundary-value problems for mixed type equations. Namely, we shall construct the regularizators for second order equations; we shall consider some new boundary-value problems for second order equations; and we shall give some examples of problems with an oblique derivative.

2.2.1. The mixed problem for second order equations not solved for the time derivative

In the papers of Lar'kin (1980, 1983) and Mamontov (1978) the following equations

$$u_{xt} - K(\ )u_{xx} - \Delta_y u = f(x, y, t) \tag{*}$$

were studied. These equations arise when we research some processes of nonstationary gas dynamics. The function K depends either on the independent variables x, y, t in the linearized case, or on the derivative u_x in the physical case. The setting of the boundary-value problems suggested in the paper cited above has one peculiarity. The authors had supposed that the two first terms from the left side of equation (*) represent a differential operator of the first order relative to the function u_x. In particular, if, in the linear case $\big(K = K(x, y, t)\big)$, equation (*) is represented in the form

$$\frac{\partial}{\partial x}\Big(u_t - K(x, y, t)u_x\Big) - \Delta_y u + \hat{K}(x, y, t)u_x = f,$$

then, as is easy to see, the boundary-value problem investigated by N. A. Lar'kin and E. V. Mamontov (the linear case, especially, was investigated in Lar'kin (1983)) coincides with the boundary-value problem I. On the other hand, equation (*) (in the linear case) may be represented as the mixed type equation (i.e. as an equation of the form (1.3.1)) and in another way:

$$Lu \equiv \frac{\partial}{\partial t}(u_x) - K(x, y, t)u_{xx} - \Delta_y u = f.$$

Then, the boundary-value problem I for these equations will vary from the boundary-value problem investigated in the papers cited above. Namely, this new boundary-value problem that will be investigated in this section is essentially a more general case.

We shall see further that, unlike the papers of N. A. Lar'kin and E. V. Mamontov, we shall not select the variable y. So, we shall consider the equation

$$Lu = lu_t - Au = f. \tag{2.2.1}$$

Here l is the first order operator

$$l = \alpha^k(x,t)\frac{\partial}{\partial x_k} + \alpha(x,t);$$

A is the second order operator

$$Au = \frac{\partial}{\partial x_i}\left(a^{ij}(x)u_{x_j}\right) - a(x)u.$$

We shall consider equation (2.2.1) in the cylinder $Q : Q = D \times (0,T)$, where D is a bounded domain in the space R^n with a smooth boundary Γ, $0 < T < +\infty$. Suppose, also, that the operator A is elliptic in the set $\bar{D}$.

Boundary-value problem: *Find solutions of equation (2.2.1) satisfying the conditions*

$$u(x,0) = u_0(x), \tag{2.2.2}$$

$$u(x,t)\Big|_{\Gamma\times(0,T)} = 0. \tag{2.2.3}$$

Establishment of solvability of boundary-value problem (2.2.1) - (2.2.3) is based on the regularization method. For $\varepsilon > 0$, we consider the operator

$$L_\varepsilon u \equiv \varepsilon\left(\frac{\partial}{\partial t} - \Delta\right)u_t + Lu$$

and the boundary value problem: *find solutions of the equation*

$$L_\varepsilon u = f, \tag{2.2.1$_\varepsilon$}$$

satisfying conditions (2.2.2), (2.2.3) and the condition

$$u_t(x,0) = u_1(x). \tag{2.2.4}$$

The function $u_1(x)$ we shall take as the solution of the equation

$$lu_1 = f(x,0) + Au_0. \tag{2.2.5}$$

Theorem 2.2.1. *Suppose the operator A is symmetric in the set $\bar{D}$ and the condition*

$$\alpha(x,t) - \frac{1}{2}\alpha^k_{x_k}(x,t) \geq \alpha_0 > 0 \quad \text{for} \quad (x,t) \in \bar{Q}$$

holds. Then, for each functions $f(x,t)$ and $u_0(x)$ such that f and f_t belong to the space $L_2(Q)$ and $u_0(x)$ and $u_1(x)$ belong to the space $W_2^2(D) \cap \overset{0}{W}{}_2^1(D)$, there exists a solution $u(x,t)$ of boundary-value problem (2.2.1)-(2.2.3) such that

$$u \in L_\infty\Big(0,T; W_2^2(D) \cap \overset{0}{W}{}_2^1(D)\Big),$$

$$u_t \in L_\infty\Big(0,T; \overset{0}{W}{}_2^1(D)\Big), \qquad u_{tt} \in L_2(Q).$$

Proof. For $\varepsilon > 0$, boundary-value problem (2.2.1$_\varepsilon$), (2.2.2)-(2.2.4) has a regular solution $u^\varepsilon(x,t)$, so that the inclusions

$$u \in L_\infty\Big(0,T; W_2^2(D) \cap \overset{0}{W}{}_2^1(D)\Big), \quad u_t \in L_\infty\Big(0,T; W_2^2(D) \cap \overset{0}{W}{}_2^1(D)\Big),$$

$$u_{tt} \in L_\infty\Big(0,T; \overset{0}{W}{}_2^1(D)\Big), \qquad u_{ttt} \in L_2(Q)$$

hold. The proof of this evident statement may be obtained, for example, by the Galerkin method. The choice of the function $u_1(x)$ denotes that for solutions of problem (2.2.1$_\varepsilon$), (2.2.2)-(2.2.4), in addition, the estimate

$$\int_D u_{tt}^2(x,0)dx \le N_1 \tag{2.2.6}$$

holds with the constant N_1 independent of ε. Taking this estimate into account, integrating by parts in the equalities

$$\int_0^t\int_D L_\varepsilon u \cdot u_\tau dx d\tau = \int_0^t\int_D f u_\tau dx d\tau,$$

$$\int_0^t\int_D \frac{\partial}{\partial \tau}\Big(L_\varepsilon u\Big) \cdot u_{\tau\tau} dx d\tau = \int_0^t\int_D f_\tau u_{\tau\tau} dx d\tau,$$

$$\int_0^t\int_D L_\varepsilon u \cdot A u_\tau dx d\tau = \int_0^t\int_D f A u_\tau dx d\tau,$$

using the hypotheses of the theorem, the second basic inequality for elliptic operators, the Young inequality, and the Gronwall inequality, we may establish the estimate

$$\varepsilon \int_D u_{tt}^{\varepsilon^2} dx + \varepsilon \sum_{i,j=1}^n \int_0^t\int_D u_{x_i x_j \tau}^{\varepsilon^2} dx d\tau + \sum_{i=1}^n \int_D \Big[u_{x_i}^{\varepsilon^2}(x,t) + u_{x_{it}}^{\varepsilon^2}(x,t)\Big] dx$$

$$+ \sum_{i,j=1}^n \int_D u_{x_i x_j}^{\varepsilon^2} dx + \int_0^t\int_D u_{\tau\tau}^{\varepsilon^2} dx d\tau \le N_2. \tag{2.2.7}$$

Here, the constant N_2 is independent of ε (it depends only on the functions $f(x,t)$, $u_0(x)$, coefficients of the operator L, the domain D, and the number T). This estimate is sufficient for passing to the limit for $\varepsilon \to 0$, and for obtaining, as a result, the desired solution of boundary-value problem (2.2.1)-(2.2.3). The theorem is proved.

Remember the definition of the function $u_1(x)$: the special choice of this function allowed us to obtain estimate (2.2.6) and, further, estimate (2.2.7). However, the function $u_1(x)$ is determined as the solution of equation (2.2.5) of the first order. At the same time, from the hypotheses of the theorem, it follows that the function $u_1(x)$ vanishes in the whole boundary of the domain D. Thus, this condition is an additional condition for the functions $f(x,t)$, $u_0(x)$, and the coefficients of the operator L.

2.2.2. Boundary-value problems for the second order equations of the mixed type

In the paper of Karatoprakliev (1977), Vragov (1977), Terekhov (1980), and Kuz'min (1985) the following equation

$$Lu \equiv K(x,t)u_{tt} - a^{ij}(x,t)u_{x_i x_j} + a^i(x,t)u_{x_i}$$

$$+a(x,t)u_t + c(x,t)u = f \qquad (**)$$

was studied. Here, the quadratic form $a^{ij}(x,t)\xi_i\xi_j$ is nonnegatively determined, and the sign of the function $K(x,t)$ is not fixed (the equation with an indeterminate characteristic form, or equation of mixed type). As is shown in Section 3.1, Chapter I, the boundary-value problem studied in the papers mentioned above coincides with the boundary-value problem I, if the equation (**) we represent as equation (1.3.1) with the operator l, where

$$l = \frac{\partial}{\partial t}.$$

The conditions of the papers mentioned above and this representation denote that the order of equation (**) with respect to the variable t may decrease in separate points no more than on a unity. The question arises, may we change the boundary-value problem and the conditions for the coefficients of equation (**) so that such degeneration holds for two units? Further, we shall show this if we use the representation of equation (**) in form (1.3.1).

We shall restrict ourselves to consideration of a certain model situation. Namely, in the cylinder Q, we consider the equation

$$Lu \equiv K(x,t)u_{tt} - \Delta u + a(x,t)u_t + c(x,t)u = f(x,t). \qquad (2.2.8)$$

Suppose, the function $K(x,t)$ is of the form $K(x,t) = K_0(x,t)\varphi(t)$. As usual, we shall assume that both the function $K_0(x,t)$, $\varphi(t)$ and the functions

$a_0(x,t)$ and $c_0(x,t)$ introduced below are sufficiently smooth. The signs of the functions $K_0(x,t)$ and $\varphi(t)$ we shall not fix. We define the functions $a_0(x,t)$ and $c_0(x,t)$ as follows

$$a_0(x,t) = \frac{a(x,t) - \lambda K_0(x,t) - K_0(x,t)\varphi'(t)}{\varphi(t)},$$

$$c_0(x,t) = c(x,t) - \lambda a_0(x,t),$$

where λ is a certain constant.

We set $l = \varphi(t)\frac{\partial}{\partial t} + \lambda$. Then, equation (2.2.8) may be written as follows:

$$Lu \equiv K_0(x,t)\frac{\partial}{\partial t}(lu) - \Delta_x u + a_0(x,t)lu + c_0(x,t)u = f.$$

Evidently, in the points where $\varphi(t) = 0$ and $K_0(x,t) = 0$ simultaneously, the order of equation (2.2.8) with respect to the variable t will decrease by two units. At the same time, recalling the setting of the boundary-value problem I$'$ for this equation, Theorem 1.3.1$'$ on uniqueness of regular solutions of the boundary-value problem I$'$, Remark 1 for Lemma 1.3.1, and the scheme of the proof of Theorem 1.3.2, we may easily show that the boundary-value problem I$'$ for equation (2.2.8) will be well-posed (the well-posedness conditions may be easily written); and this problem will have an almost regular solution. It is clear that a situation is possible that the bases of the cylinder Q will be free from the boundary conditions, or from condition (1.3.2$'$). At the same time, the papers cited above mean the value of the function $u(x,t)$ be given in one of the bases of the cylinder Q.

2.2.3. The problem with oblique derivative for equations of second and third orders

The problems with oblique derivative, i.e. the problems when the values of a certain differential operator of first order are given in the boundary of the domain or in the part of it, are studied sufficiently well. We may cite the following monographs: Ladyzhenskaya *et al.*, 1967; Ladyzhenskaya and Ural'tseva, 1973; Blokhin, 1995; the papers of Kreiss, 1970; Sakamoto, 1970; Bizhanova and Solonnikov, 1993; and many others. However, the methods of investigation of solvability of these problems are classical. This means that they are based on the a *priori* estimates, integral representations, and the imbedding theorems. Vragov (1987) had suggested the idea of passing to the mixed type equation when considering the problem with oblique derivative for hyperbolic equations in the case of constant coefficients and in the case of the special domain (Vragov, 1987, 1989). In some cases this idea may be used when investigating the problems with oblique derivative for parabolic and elliptic equations with variable coefficients in the domains of sufficiently arbitrary geometry. However, since the problem with oblique derivative for such equations is well-studied, we shall state the idea of using the theory of mixed type equations taking a new problem - the problem with oblique derivative for the third order equations, the equations of type (1.3.23).

The idea of using the theory of mixed type equations denotes that we must pass from the initial equation to the equation of the order higher than the initial one on the unity. As we shall study the problem with oblique derivative for the third order equations, then we must study the solvability of the boundary-value problem for fourth order equations.

So, we shall consider the equation

$$Lu = L_0 lu + L_1 u = f(x,t) \tag{2.2.9}$$

in the cylinder Q.

Here

$$L_0 u = u_{tt} - Au_t - Bu,$$

$$Au = \frac{\partial}{\partial x_i}\left(a^{ij}(x)u_{x_j}\right) - a_0(x)u,$$

$$Bu = \frac{\partial}{\partial x_i}\left(b^{ij}(x)u_{x_j}\right) - b_0(x)u,$$

$$L_1 u = A_1 u_t + B_1 u,$$

$$A_1 u = a^{1,ij}(x)u_{x_i x_j} + a^{1,i}(x)u_{x_i} + a^1(x)u,$$

$$B_1 u = b^{1,ij}(x)u_{x_i x_j} + b^{1,i}(x)u_{x_i} + b^1(x)u,$$

$$lu = u_t + \alpha^k(x)u_{x_k} + \alpha_0(x)u.$$

Equation (2.2.9) is a twice composite equation. Such equations we have not studied yet. However, the general idea of solvability research for boundary-value problems for mixed type equations stated in Chapter 1 will allow us to research solvability of some boundary-value problems for these equations. We confine ourselves to presentation only of those results on solvability of the boundary-value problems for equation (2.2.9) which we shall need further.

We denote by

$$\kappa(x) = \alpha^k(x)\nu_k(x)$$

the function determined for $x \in \Gamma$. We suppose that the following condition holds.

Condition (l). *The function* $\kappa(x)$ *is nonpositive.*

Under this supposition we consider the following boundary-value problem: *find a solution of equation (2.2.9) which satisfies the conditions*

$$u(x,0) = u_0(x), \quad u_t(x,0) = u_1(x), \quad u_{tt}(x,0) = u_2(x), \tag{2.2.10}$$

$$lu\Big|_{\Gamma\times(0,T)} = 0. \tag{2.2.11}$$

We shall introduce the following notations

$$v_0(x) = u_1(x) + \alpha^k(x)u_{0x_k}(x) + \alpha_0(x)u_0(x),$$

$$v_1(x) = u_2(x) + \alpha^k(x)u_{1x_k}(x) + \alpha_0(x)u_1(x).$$

Theorem 2.2.2. *Suppose the condition* (l) *holds, and the operator* A *is a symmetric elliptic operator for* $x \in \bar{D}$. *If the functions* $f(x,t)$, $u_0(x)$, $u_1(x)$, *and* $u_2(x)$ *are such that* $f(x,t)$ *belongs to the space* $L_2(Q)$; $u_0(x)$, $v_0(x)$, *and* $u_1(x)$ *belong to the space* $W_2^2(D)$; $u_2(x)$ *and* $v_1(x)$ *belong to the space* $W_2^1(D)$; $u_0(x)$, $v_0(x)$, *and* $v_1(x)$ *vanish in* Γ, *then there exists a function* $u(x,t)$ *such that* u *and* u_t *belong to the space* $L_\infty\left(0,T;W_2^2(D)\right)$;

lu belongs to the space $L_\infty\Big(0,T;W_2^2(D)\cap \overset{0}{W}{}_2^1(D)\Big)$; $(lu)_t$ belongs to the space $L_2\Big(0,T;W_2^2(D)\Big)\cap L_\infty\Big(0,T;\overset{0}{W}{}_2^1(D)\Big)$; and $(lu)_{tt}$ belongs to the space $L_2(Q)$. This function $u(x,t)$ is a solution of equation (2.2.9) satisfying initial and boundary conditions (2.2.10) and (2.2.11).

Proof. We shall use the method of continuation by parameter. To this end, we consider the family of the operators

$$L^\lambda u \equiv L_0 lu + \lambda L_1 u, \qquad \lambda \in [0,1],$$

and the family of the boundary-value problems: *find the solutions of the equation*

$$L^\lambda u = f(x,t), \tag{2.2.9$_\lambda$}$$

satisfying conditions (2.2.10) and (2.2.11).

For $\lambda = 0$, problem (2.2.9$_0$), (2.2.10), (2.2.11) is the usual initial boundary-value problem for the function $v = lu$ for a particular case of equation (1.3.23):

$$L^0 u \equiv L_0 v = f(x,t),$$

$$v(x,0) = v_0(x), \quad v_t(x,0) = v_1(x), \quad v\Big|_{\Gamma\times(0,T)} = 0.$$

If the hypotheses of the theorem hold, this problem will be solvable for the function v. In this case, the solution $v(x,t)$ will satisfy the following inclusions

$$v \in L_\infty\Big(0,T;W_2^2(D)\cap \overset{0}{W}{}_2^1(D)\Big),$$

$$v_t \in L_2\Big(0,T;W_2^2(D)\Big)\cap L_\infty\Big(0,T;\overset{0}{W}{}_2^1(D)\Big), \qquad v_{tt} \in L_2(Q).$$

This statement may be easily established by the methods of Chapter 1. Now, if we know the function $v(x,t)$, we may easily find the function $u(x,t)$ as the solution of the problem

$$lu = v, \qquad u(x,0) = u_0(x). \tag{2.2.12}$$

In this case, the function $u(x,t)$ will satisfy the inequalities:

$$\left\|u(x,t)\right\|^2_{W_2^k(D)}(t) \leq M\left(\int_0^t \left\|v\right\|^2_{W_2^k(D)}(\tau)d\tau + \left\|u_0\right\|^2_{W_2^2(D)}\right),$$

$$\left\|u_t(x,t)\right\|^2_{W_2^k(D)}(t) \leq M\left(\int_0^t \left\|v\right\|^2_{W_2^k(D)}(\tau)d\tau + \left\|u_1\right\|^2_{W_2^2(D)}\right), \tag{2.2.13}$$

where $k = 1, 2$; $t \in [0, T]$, and the constant M depends on the coefficients of the operator l, the domain D, and the number T. Solvability of problem (2.2.12) follows from Lemma 1.3.1. The proof of estimates (2.2.13) is analogous to those of Lemma 1.3.1 but with use of the Gronwall lemma.

Thus, for $\lambda = 0$, there exists the solution of problem ($2.2.9_0$), (2.2.10), (2.2.11) from the desired class. In order to justify the methods of continuation in a parameter we need to obtain the estimates of solutions of problem ($2.2.9_\lambda$), (2.2.10), (2.2.11) in the desired class of functions uniform with respect to λ.

The methods of establishment of these estimates are just standard.

The equality

$$\int_0^t\int_D L^\lambda u \cdot v_\tau dx d\tau = \int_0^t\int_D f v_\tau dx d\tau \qquad (v = lu)$$

after integrating by parts and applying the Young inequality and the Gronwall lemma gives the inequality

$$\int_D v_t^2(x,t)dx + \int_0^t \left\|v_\tau\right\|^2_{W_2^1(D)}(\tau)d\tau$$

$$\leq M_1 \left[\int_0^t \left(\left\| u_\tau \right\|^2_{W_2^1(D)}(\tau) + \left\| u \right\|^2_{W_2^1(D)}(\tau) \right) d\tau + 1 \right]. \qquad (2.2.14)$$

Here $t \in [0, T]$, the constant M_1 depends only on the coefficients of the operator L, the domain D, the number T, and the functions $f(x,t), u_0(x)$, and $u_1(x)$.

Secondly, we consider the equality

$$\int_0^t \int_D L^\lambda u \cdot Av dx d\tau = \int_0^t \int_D f \cdot Av dx d\tau.$$

Again, integrating by parts, using the Young inequality, the second basic inequality for elliptic operators, inequality (2.2.14), and the Gronwall lemma, we obtain the inequality

$$\left\| v \right\|^2_{W_2^2(D)}(t) \leq M_2 \left[\int_0^t \left(\left\| u_\tau \right\|^2_{W_2^2(D)}(\tau) + \left\| u \right\|^2_{W_2^2(D)}(\tau) \right) d\tau + 1 \right]. \qquad (2.2.15)$$

Here $t \in [0, T]$, and the constant M_2 depends only on the coefficients of the operator L, the domain D, the number T, and the functions $f(x,t), u_0(x)$, and $u_1(x)$.

The inequality

$$\int_0^t \int_D L^\lambda u \cdot Av_\tau dx d\tau = \int_0^t \int_D f \cdot Av_\tau dx d\tau$$

we analyze again in a standard way. After using the second basic inequality for elliptic operators and inequality (2.2.15), we obtain the inequality

$$\left\| v_t \right\|^2_{W_2^1(D)}(t) + \int_0^t \left\| v_\tau \right\|^2_{W_2^2(D)}(\tau) d\tau$$

$$\leq M_3 \left[\int_0^t \left(\left\| u_\tau \right\|^2_{W_2^2(D)}(\tau) + \left\| u \right\|^2_{W_2^2(D)}(\tau) \right) d\tau + 1 \right]. \qquad (2.2.16)$$

Here $t \in [0, T]$, and the constant M_3 depends on similar values as the constants M_1 and M_2.

The following evident inequalities

$$\left\| u \right\|^2_{W_2^2(D)}(\tau) \leq M_4 \left[\int_0^\tau \left\| u_\xi \right\|^2_{W_2^2(D)}(\xi) d\xi + 1 \right]$$

$$\leq M_4 \left[\int_0^t \left\| u_\tau \right\|^2_{W_2^2(D)}(\tau) d\tau + 1 \right]$$

hold true. Here $0 \leq \tau \leq t \leq T$, and the constant M_4 depends only on the function $u_0(x)$ and the number T. Therefore, inequality (2.2.16) yields the inequality

$$\left\| v_t \right\|^2_{W_2^1(D)}(t) + \int_0^t \left\| v_\tau \right\|^2_{W_2^2(D)}(\tau) d\tau \leq M_5 \left[\int_0^t \left\| u_\tau \right\|^2_{W_2^2(D)}(\tau) d\tau + 1 \right]. \qquad (2.2.17)$$

To estimate the integrand term in the right side of inequality (2.2.17), we use now the second inequality (2.2.13) for $k = 2$. As a result we obtain the inequality

$$\left\| v_t \right\|^2_{W_2^1(D)}(t) + \int_0^t \left\| v_\tau \right\|^2_{W_2^2(D)}(\tau) d\tau \leq M_6 \left[\int_0^t \int_0^\tau \left\| v_\xi \right\|^2_{W_2^2(D)}(\xi) d\xi + 1 \right].$$

Applying the Gronwall lemma, we obtain the uniform boundedness of the function v_t in the space $L_2\left(0, T; W_2^2(D)\right)$. From the estimate v_t in this space and from inequalities (2.2.13)-(2.2.17), we obtain uniform boundedness of the functions $u(x,t)$ in the space $L_\infty\left(0, T; W_2^2(D)\right)$, the functions

$u_t(x,t)$ in the space $L_\infty\Big(0,T;W_2^2(D)\Big)$, the functions $v(x,t)$ in the space $L_\infty\Big(0,T;W_2^2(D)\cap\overset{0}{W}{}_2^1(D)\Big)$, and the functions $v_t'(x,t)$ in the space $L_\infty\Big(0,T;\overset{0}{W}{}_2^1(D)\Big)$. The estimate of the function v_{tt} in the space $L_2(Q)$ follows from the previous estimates.

So, we have established uniform boundedness with respect to λ, for $\lambda\in[0,1]$, of solutions of problem (2.2.9$_\lambda$), (2.2.10), (2.2.11) in the desired norms. Moreover, these constants will depend on the function $f(x,t)$ norm in the space $L_2(Q)$ and on the norms of the functions $u_0(x)$, $u_1(x)$, and $u_2(x)$ corresponding to equation (2.2.9). In other words, boundary-value problem (2.2.9$_\lambda$), (2.2.10), (2.2.11) does not have the property of smoothness decreasing. Therefore, the method of continuation in a parameter is applicable, and we have solvability of problem (2.2.9$_\lambda$), (2.2.10), (2.2.11) in the desired class for all $\lambda\in[0,1]$; in particular, for $\lambda=1$. This proves Theorem 2.2.2.

Now, let l be as follows

$$lu\equiv l_0u=\alpha^k(x)u_{x_k}+\alpha_0(x)u.$$

Evidently, in this case, for the equation

$$Lu\equiv L_0l_0u+L_1u=f(x,t) \tag{2.2.9$'$}$$

it is necessary to set the problem with the only initial conditions

$$u(x,0)=u_0(x),\qquad u_t(x,0)=u_1(x), \tag{2.2.10$'$}$$

and with the boundary condition

$$l_0u\Big|_{\Gamma\times(0,T)}=0. \tag{2.2.11$'$}$$

Denote

$$\tilde v_0(x)=\alpha^k(x)u_{0x_k}(x)+\alpha_0(x)u_0(x),$$

$$\tilde v_1(x)=\alpha^k(x)u_{1x_k}(x)+\alpha_0(x)u_1(x).$$

Theorem 2.2.3. *Let the condition* (l) *hold and suppose the operator* A *is a symmetric elliptic operator in* $\bar{D}$. *Suppose, in addition, that*

$$\alpha_0(x) \geq \bar{\alpha}_0 > 0 \quad \text{for} \quad x \in \bar{D}.$$

If the function $f(x,t)$ *belongs to the space* $L_2(Q)$, *the functions* $u_0(x)$, $u_1(x)$, *and* $\tilde{v}_0(x)$ *belong to the space* $W_2^2(D)$, *the function* $\tilde{v}_1(x)$ *belongs to the space* $W_2^1(D)$, *the functions* $u_0(x)$, $\tilde{v}_0(x)$, *and* $\tilde{v}_1(x)$ *vanish in* Γ, *and* $\bar{\alpha}_0$ *is a constant sufficiently large, then there exists a function* $u(x,t)$ *such that* u *and* u_t *belong to the space* $L_\infty\Big(0,T;W_2^2(D)\Big)$, $l_0 u$ *belongs to the space* $L_\infty\Big(0,T;W_2^2(D)\cap \overset{0}{W}{}_2^1(D)\Big)$, $(l_0u)_t$ *belongs to the space* $L_2\Big(0,T;W_2^2(D)\Big)\cap L_\infty\Big(0,T;\overset{0}{W}{}_2^1(D)\Big)$, $(l_0u)_{tt}$ *belongs to the space* $L_2(Q)$. *This function* $u(x,t)$ *solves equation (2.2.9′) and attains initial and boundary conditions (2.2.10′) and (2.2.11′).*

Proof. The proof of this theorem is just similar to the proof of Theorem 2.2.2. We should only take into account that, by Lemma 1.3.1 (the remark to this lemma), the boundary-value problem

$$l_0 u = v$$

(the problem on the cut, i.e. in the set $D\times t$, t is fixed) has a regular solution. This solution satisfies the estimates

$$\Big\|u(x,t)\Big\|^2_{W_2^2(D)} \leq M\Big\|v(x,t)\Big\|^2_{W_2^2(D)},$$

$$\Big\|u_t(x,t)\Big\|^2_{W_2^2(D)} \leq M\Big\|v(x,t)\Big\|^2_{W_2^2(D)},$$

where $t\in[0,T]$, and the constant M depends only on the coefficients of the operator l_0 and on the domain D. Moreover, both when we establish these estimates and when we establish the estimates of solutions of corresponding problems (2.2.9$_\lambda$), (2.2.10′), (2.2.11′) we have to use the condition for the value I_0 instead of the Gronwall lemma.

Now, we shall formulate one more theorem on behavior of solutions of problem (2.2.9)-(2.2.11).

Theorem 2.2.4. *Suppose the conditions of Theorem 2.2.2 hold and the monotone increasing function $\varphi(t)$ is continuously differentiable and such that*

$$\varphi(0) = 1, \quad \varphi(t) - \gamma_0\varphi'(t) \geq 0, \quad \gamma_0 > 0.$$

Suppose, also, that the following conditions hold

$$b^{ij}(x)\xi_i\xi_j \geq \lambda_0|\xi|^2, \qquad -a_0(x) \geq \bar{a}_0,$$

$$-b_0(x) \geq \bar{b}_0, \quad \alpha_0(x) \geq \bar{\alpha}_0, \quad -c^{ij}(x)\xi_i\xi_j \geq 0, \quad c^{ij}(x) = c^{ji}(x),$$

$$\left[2\alpha_0(x)c^{ij}(x) - \left(\alpha^k(x)c^{ij}(x)\right)_{x_k} + 2\alpha^j_{x_k}c^{jk}(x)\right]\xi_i\xi_j \leq 0,$$

where $x \in \bar{D}$, $\xi \in R^n$, $\bar{a}_0$, $\bar{b}_0$, and $\bar{\alpha}_0$ are positive constants sufficiently large, *and the function $f(x,t)$ is such that*

$$\int_0^\infty\int_D \varphi(t)f^2(x,t)\,dxdt < +\infty.$$

Then, for the solutions of problem (2.2.9)-(2.2.11) the following estimate

$$\left\|v_t(x,t)\right\|^2_{L_2(D)}(t) + \left\|v(x,t)\right\|^2_{W_2^1(D)}(t)$$

$$+\left\|u_t(x,t)\right\|^2_{L_2(D)}(t) + \left\|u(x,t)\right\|^2_{W_2^1(D)}(t) \leq \frac{M_0}{\varphi(t)}$$

will hold. Here $t \geq 0$, and the constant M_0 depends only on the coefficients of the operator L and the functions $f(x,t)$, $u_0(x)$, $u_1(x)$.

Proof. The proof of this theorem is based on analysis of the equality

$$\int_0^t\int_D \varphi(t)Lu\left(v_\tau + \rho v\right)dxd\tau = \int_0^t\int_D \varphi(\tau)f\left(v_t + \rho v\right)dxd\tau.$$

This analysis is tedious but sufficiently standard. Here $\rho > 0$ is sufficiently small. In this case, parallel with the hypotheses of the theorem, we use the inequalities

$$\int_0^t\int_D \varphi(\tau)\left[u^2+\sum_{i=1}^n u_{x_i}^2\right]dxd\tau \le K_1\left(\int_0^t\int_D \varphi(\tau)\left[v^2+\sum_{i=1}^n v_{x_i}^2\right]dxd\tau+1\right),$$

$$\int_0^t\int_D \varphi(\tau)\left[u_\tau^2+\sum_{i=1}^n u_{x_i\tau}^2\right]dxd\tau \le K_1\left(\int_0^t\int_D \varphi(\tau)\left[v_\tau^2+\sum_{i=1}^n v_{x_i\tau}^2\right]dxd\tau+1\right),$$

which are deduced by means of the condition imposed on the constant $\bar{\alpha}_0$ if we analyze the equalities

$$\int_0^t \varphi(\tau)\Big(lu,u\Big)_{W_2^1(D)}(\tau)d\tau = \int_0^t \varphi(\tau)\Big(v,u\Big)_{W_2^1(D)}(\tau)d\tau,$$

$$\int_0^t \varphi(\tau)\Big(lu_\tau,u_\tau\Big)_{W_2^1(D)}(\tau)d\tau = \int_0^t \varphi(\tau)\Big(v_\tau,u_\tau\Big)_{W_2^1(D)}(\tau)d\tau$$

Here $(,)_{W_2^1(D)}(\tau)$ is the scalar product in the space $W_2^1(D)$ in the cut $t=\tau$.

Now, we shall investigate directly the problem with oblique derivative. Namely, we shall consider, at first, the problem: *find a solution of equation*

$$L_0u = g(x,t), \tag{2.2.18}$$

which satisfies conditions (2.2.10′) and (2.2.11).

We set $u_2(x) = Au_1(x) + Bu_0(x) + g(x,0)$ and suppose the functions $v_0(x)$ and $v_1(x)$ are defined through the functions $u_0(x)$, $u_1(x)$ and $v_2(x)$ as before.

Theorem 2.2.5. *Let the condition* (l) *hold, and the operator* A *be a symmetric elliptic operator in* $\bar{D}$. *Suppose the function* $g(x,t)$ *belongs to the space* $L_2\Big(0,T;W_2^1(D)\Big)$; *the function* $g_t(x,t)$ *belongs to the space* $L_\infty\Big(0,T;L_2(D)\Big)$; *the functions* $u_0(x)$, $u_1(x)$, *and* $v_0(x)$ *belong to the space* $W_2^2(D)$; *the functions* $u_2(x)$ *and* $v_1(x)$ *belong to the space* $W_2^1(D)$; *the functions* $u_0(x)$, $v_0(x)$, *and* $v_1(x)$ *vanish in* Γ. *Then there exists the function* $u(x,t)$ *such that* u *and* u_t *belong to the space* $L_\infty\Big(0,T;W_1^2(D)\Big)$; lu *belongs to the space* $L_\infty\Big(0,T;W_2^2(D)\cap \overset{0}{W}{}_2^1(D)\Big)$; *and* u_{tt} *belongs to the space* $L_\infty\Big(0,T;L_2(D)\Big)$. *This function* $u(x,t)$ *is a solution of equation (2.2.18) and attains initial and boundary values (2.2.10′) and (2.2.11).*

Proof. Under the conditions of Theorem 2.2.5, there exists the function $u(x,t)$ which solves problem (2.2.9)-(2.2.11) with the operators A and B_1 and the function f, which will be pointed out below, from the class mentioned in the formulation of Theorem 2.2.2.

For the function $u(x,t)$, the following inclusions

$$lAu_t \in L_2(Q), \qquad lBu \in L_2(Q) \tag{2.2.19}$$

hold. Really, for the smooth functions, we have

$$lAu_t = Alu_t - \Phi_A u_t, \qquad lBu = Blu - \Phi_B u,$$

where Φ_A and Φ_B are certain operators of the second order with respect to the variables $x_1,\dots,x_n$. As the function $u(x,t)$ is such that the right sides of these equalities belong to the space $L_2(Q)$, the left sides will belong to the same space also. Closing these equalities we obtain inclusions (2.2.19).

Further, the inclusion $(lu)_{tt} \in L_2(Q)$ immediately follows from the fact that $u(x,t)$ belongs to the class pointed out in Theorem 2.2.2. Together with inclusions (2.2.18) we obtain that the inclusion

$$l\Big(u_{tt} - Au_t - Bu\Big) \in L_2(Q)$$

holds.

We shall solve problem (2.2.9)-(2.2.11) with the special right side: $f(x,t) = lg(x,t)$. In this case, we shall have the equality

$$l\Big(u_{tt} - Au_t - Bu - g\Big) + \Big(A_1 - \Phi_{A}\Big)u_t + \Big(B_1 - \Phi_{B}\Big)u = 0. \qquad (2.2.20)$$

We suppose now that the operators A_1 and B_1 are chosen so: $A_1 = \Phi_{A}$, $B_1 = \Phi_{B}$. Then, equality (2.2.20) will take a form

$$l\Big(u_{tt} - Au_t - Bu - g\Big) = 0.$$

The following equality

$$u_{tt} - Au_t - Bu - g\Big|_{t=0} = 0$$

holds. So, we have obtained for the function $w = u_{tt} - Au_t - Bu - g$ that

$$lw = 0, \qquad w(x,0) = 0.$$

Taking into account the condition (l), we obtain

$$w(x,t) \equiv 0.$$

So, the function $u(x,t)$ constructed as a solution of problem (2.2.9)-(2.2.11) will be a solution of problem (2.2.18), (2.2.10′), (2.2.11) with oblique derivative.

The fact that the function $u(x,t)$ belongs to the class pointed out in Theorem 2.2.2 denotes that this function will belong to the class desired in Theorem 2.2.5. The theorem is proved.

The problem with oblique derivative for equation (2.2.18), when instead of condition (2.2.11) we take condition (2.2.11′), is stated similarly. In this case, we must set only one initial condition, and the function $u_1(x)$ may be taken specially. In order to show that the solution of problem (2.2.9′), (2.2.10′), (2.2.11′) will be a solution of the problem with oblique derivative set above, it suffices to prove that, for almost all $t \in [0,T]$, in the cuts of the cylinder Q, we have

$$l_0\Big(u_{tt} - Au_t - Bu - g\Big) = 0.$$

To this end, we need that the inclusions of Theorem 2.2.3 will be valid for the function u_t. These inclusions will hold if, in the conditions of Theorem 2.2.3, we require that the function f_t belongs to the space $L_2(Q)$ (then we may pass to the equation (2.2.9′) differentiated with respect to t and obtain the inclusions for the function u_t established in Theorem 2.2.3).

Theorem 2.2.4 in application to the problem with oblique derivative allows us to describe the behavior of solution of this problem when $t \to \infty$.

Theorem 2.2.6. *Let the conditions of Theorem 2.2.4 hold for the functions $\varphi(t)$, $b^{ij}(x)$, $a_0(x)$, $b_0(x)$, $\alpha_0(x)$, and the functions*

$$c^{ij}(x) = 2\alpha^i_{x_k}(x)a^{kj}(x) - \alpha^k(x)a^{ij}_{x_k}(x).$$

If the function $g(x,t)$ satisfies the inequality

$$\int_0^\infty\int_D \varphi(t)\Big[lg(x,t)\Big]^2 dxdt < \infty,$$

then solutions of problem (2.2.18), (2.2.10′), (2.2.11) satisfy the estimate of Theorem 2.2.4.

Proof. The proof of this theorem is evident. The scheme stated above allows us to establish solvability of the problem with oblique derivative, to investigate the properties of its solutions in some other cases. So, it easy to prove solvability of the boundary-value problems for equation (2.2.17) when we set the values lu or l_0u in the lateral area of the cylinder Q, and the conditions of periodic type with respect to the variable t

$$u(x,0) = \gamma_0 u(x,T) + u_0(x),$$

$$u_t(x,T) = \gamma_1 u_t(x,T) + u_1(x)$$

(γ_0 and γ_1 are constants).

Similarly, we may investigate solvability of the problem with oblique derivative for the first order equations with respect to t and the operator L_0 being as follows

$$L_0 u = u_t - Au_t - Bu.$$

This method is applicable also to elliptic equations, to equations of type (1.3.52), and to some other equations. Undoubtedly, our conditions of solvability of the problem with oblique derivative require the smoother entering data than the classical conditions mentioned, for example, by Ladyzhenskaya *et al.*, 1967; Ladyzhenskaya and Ural'tseva, 1973; and by Blokhin, 1995. However, simplicity of establishment and the possibility of wide applications makes this method valuable.

Remark. The (l)-condition is one of the basic conditions of solvability of the problem with oblique derivative. This condition arises in practically all the works devoted to problems of solvability of the problems with oblique derivative.

2.2.4. A problem of viscous elasticity and the third order equation in noncylindrical domains connected with this problem

In Cannarsa *et al.*, (1990) one problem of viscous elasticity was investigated. Its mathematical statement is as follows.

Let Q_T be a domain of the variables (x, t), $0 < t < T < +\infty$ such that for each $t^* \subset (0, T)$ the element x is an element from the domain D_t from the space R^n lying in the hyperplane $t = t^*$. In other words, Q_T is a noncylindrical domain

$$Q_T = \bigcup_{0<t<T} \Big(D_t \times \{t\}\Big)$$

with the lower basis D_0 and the upper basis D_T.

The viscous elasticity problem: *find solution of the strongly damped wave equation*

$$u_{tt} = \Delta(u + ku_t) + f(x, t) \quad in \quad Q_T \tag{2.2.21}$$

($k > 0$ is a constant), satisfying the conditions

$$u(x,0) = u_0(x), \qquad u_t(x,0) = u_1(x), \qquad x \in D_0, \tag{2.2.22}$$

$$u(x,t) + ku_t(x,t) = 0 \quad \text{in} \quad S_T = \bigcup \Big(\partial D_t \times \{t\}\Big). \tag{2.2.23}$$

In the paper cited above, for problem (2.2.21)-(2.2.23), the existence of the generalized solution was established. Using the results on solvability of boundary-value problems for mixed type equations stated in Chapter I, we may prove existence of regular solutions in some cases both for problem (2.2.21)-(2.2.23) and for the modified problem. We shall remark the following for modified problem (2.2.21). For the generalized solution, the existence of which was established by Cannarsa *et al.*, (1990), the boundary conditions are false (generally speaking). Thus, the generalized solution does not take into account the form of Q_T. Namely, it does not take into account whether the domain Q_T is extending or contracting. If we seek such a solution which satisfies the boundary conditions, we need to define the conditions for the domain Q_T more correctly. Moreover, such a situation may occur that boundary condition (2.2.23) will be insufficient for existence of the regular solution.

The research of the viscous elasticity problem is a simple one-dimensional case.

Suppose Q_T is such that

$$Q_T = \Big\{(x,t): \ \ 0 < t < T < +\infty, \ \ 0 < x < \alpha(t)\Big\},$$

where $\alpha(t)$ is a smooth function, positive in $[0,T]$ and such that $\alpha(0) = 1$. For simplicity, we consider the case $k = 1$. We introduce a new variable $z : z = x/\alpha(t)$. Evidently, the point (z,t) now belongs to the cylinder $Q = \{(z,t) : 0 < z < 1, \ 0 < t < T\}$. Equation (2.2.21) will take the following form in the variables z and t

$$\left(\frac{\partial}{\partial t} - \frac{1}{\alpha^2(t)}\frac{\partial^2}{\partial z^2} - \frac{z\alpha'(t)}{\alpha(t)}\frac{\partial}{\partial z} - 1\right)\left(v_t - \frac{z\alpha'(t)}{\alpha(t)}v_z + v\right) + v = \tilde{f}(z,t), \tag{2.2.21$'$}$$

where

$$v(z,t) = u\Big(\alpha(t)z, t\Big), \qquad \tilde{f}(z,t) = f\Big(\alpha(t)z, t\Big).$$

Now, we introduce the following notations

$$A = \frac{\partial}{\partial t} - \frac{1}{\alpha^2(t)} \frac{\partial^2}{\partial z^2} - \frac{z\alpha'(t)}{\alpha(t)} \frac{\partial}{\partial z} - 1,$$

$$l = \frac{\partial}{\partial t} - \frac{z\alpha'(t)}{\alpha(t)} \frac{\partial}{\partial z} + 1,$$

$$B = I.$$

Then, equation (2.2.21′) will become as follows

$$Alv + Bv = \tilde{f},$$

i.e. to within the limits of transposition of the operators A and l, it becomes as (1.3.1). As is shown in §3, Chapter 1, the well-posed boundary-value problem for equation (2.2.20′) is defined, first of all, by the coefficients of the operator l. If we return back to equation (2.2.20), we may say that the well-posed problem for this equation is defined by the function $\alpha(t)$.

Now, we shall formulate the exact result for equation (2.2.20).

Let D_τ is the interval $D_\tau = \{(x,t) : t = \tau,\ \ 0 < x < \alpha(\tau)\}$.

Theorem 2.2.7. *Suppose that*

$$\alpha(t) \geq \alpha_0 > 0, \quad \alpha'(t) < 0, \quad \textit{for} \quad t \in [0,T].$$

Suppose also that a function $f(x,t)$ belongs to the space $L_2(Q_T)$; a function $u_0(x)$ belongs to the space $W_2^1([0,1])$ and vanishes for $x = 0$; a function $u_1(x)$ belongs to the space $W_2^1([0,1])$; the function $u_0(x) + u_1(x)$ vanishes for $x = 0$, $x = 1$. Then, there exists such a function $u(x,t)$ which solves equation (2.2.21); $u(x,t)$ belongs to the space $W_2^1(D_t)$ for almost all t from the interval $[0,T]$; $u_t(x,t)$ belongs to the space $L_2(D_t)$ for almost all t from $[0,T]$; $u_t(x,t) + u(x,t)$ belongs to the space $W_2^{2,1}(Q_T)$. Moreover, the function $u(x,t)$ satisfies initial conditions (2.2.22) and the boundary conditions

$$u\Big|_{x=0} = 0, \qquad u + u_t\Big|_{x=\alpha(t)} = 0. \tag{2.2.24}$$

If

$$\alpha(t) \geq \alpha_0 > 0, \quad \alpha'(t) > 0 \quad \textit{for} \quad t \in [0, T];$$

the function $f(x,t)$ belongs to the space $L_2(Q_T)$; and the functions $u_0(x)$ and $u_1(x)$ belong to the space $\overset{0}{W}{}_2^1([0,1])$; then there exists a function $u(x,t)$ which solves equation (2.2.21) and such that $u(x,t)$ and $u_t(x,t) + u(x,t)$ belong to the space $\overset{0}{W}{}_2^1(D_t)$ for almost all t from $[0,T]$; $u_t(x,t)$ belongs to the space $\overset{0}{W}{}_2^1(D_t)$ for almost all t from $[0,T]$; $u_t(x,t)+u(x,t)$ belongs to the space $W_2^{2,1}(Q_T)$. Moreover, the function $u(x,t)$ satisfies initial conditions (2.2.22) and the boundary conditions

$$u\Big|_{x=0} = u\Big|_{x=\alpha(t)} = u + u_t\Big|_{x=\alpha(t)} = 0. \tag{2.2.25}$$

Proof. To prove this theorem, we need to use the fact that the corresponding boundary-value problem for equation (2.2.21′) has a regular solution. Then we return to equation (2.2.21) (i.e. we pass from the variables (z,t) to the variables (x,t)).

If we compare Theorem 2.2.7 and the result from Cannarsa *et al.* (1990), we see that in the last case the problem was considered satisfying the condition $\alpha'(t) < 0$ only.

If the function $\alpha'(t)$ changes sign in the interval $(0,T)$, we need to consider the boundary-value problem with the combination of conditions (2.2.24) and (2.2.25). Suppose a domain Q_T is a domain with two curvilinear boundaries

$$Q_T = \Big\{(x,t): \ 0 < t < T < +\infty, \ \alpha_1(t) < x < \alpha_2(t)\Big\},$$

where $\alpha_2(t) > \alpha_1(t)$. Then, making the change

$$z = \frac{x - \alpha_1(t)}{\alpha_2(t) - \alpha_1(t)}$$

we reduce the domain Q_T to the cylindrical domain again, and equation

(2.2.21) becomes an equation of the mixed type. Using the theory of well-posed boundary-value problems for equations of mixed type, we may establish existence or regular solutions of corresponding boundary-value problems.

In the multidimensional case, the following case may be researched analogously. Suppose D_0 is a bounded domain from R^n with a smooth boundary and let $\alpha_1(t), \alpha_2(t), \ldots, \alpha_n(t)$ be positive and smooth for $t \in [0, T]$ and $\alpha_i(0) = 1$. The domain D_t we define as follows

$$D_t = \left\{ x = (x_1, \ldots, x_n) : \left(\frac{x_1}{\alpha_1(t)}, \ldots, \frac{x_n}{\alpha_n(t)} \right) \in D_0 \right\}$$

and the domain

$$Q_T = \bigcup_{0<t<T} \Big(D_t \times \{t\} \Big).$$

Making the change $z_i = x_i/\alpha_i(t)$, $i = 1, \ldots, n$, we reduce multidimensional equation (2.2.21) to the equation of composite type in a cylindrical domain. Using the results on existence of regular solutions of the boundary-value problems for equations of composite type, we may (again) obtain existence of regular solutions of corresponding boundary-value problems for equation (2.2.21).

Conclusion

The results stated in the monograph, of course, do not give the complete representation on development of the theory of composite type equations and its applications. We do not discuss, for example, the nonlinear equations of composite type. Also, we do not give the results on qualitative properties of solutions (smoothness, behavior of solution depending on the parameter, and so on). Also we do not set the problems of solvability of boundary-value problems for operators whose higher derivative involves superposition of hyperbolic and hyperbolic-parabolic operators. On the one hand, we do not discuss these problems because the author does not have results of his own in this area; but furthermore these results stand outside the main goal of the monograph. However, the author hopes that the questions mentioned above will be investigated and stated in further papers and monographs.

Bibliography

Agmon, S., Duglis, A., and Nirenberg, L. (1962). *Estimates Near the Solution Boundary of Elliptic Equations with Partial Derivatives for General Boundary Conditions.* Izd. Gos. Tech. Lit., Moscow (in Russian).

Akhmatov, S. S. and Belov, Yu. Ya. (1991). On some inverse problems for parabolic equations. *Sov. Math. Dokl.,* **43**(1), 166–170.

Anikonov, Yu. E. (1978). *Some Methods of Investigation of Multidimensional Inverse Problems.* Nauka, Novosibirsk (in Russian).

Anikonov, Yu. E. (1985). Some questions of the theory of inverse problems for kinetic equations. *Inverse Problems of Mathematical Physics.* Computing Center, Siberian Branch of the USSR Acad. Sci., Novosibirsk, 28–41 (in Russian).

Anikonov, Yu. E. and Bubnov, B. A. (1987). Control and inverse problem. *Sov. Math. Dokl.,* **34,** 13–14.

Anikonov, Yu. E. and Bubnov, B. A. (1988). Existence and uniqueness of a solution to an inverse problem for a parabolic equation. *Sov. Math. Dokl.,* **37,** 130–132.

Anikonov, Yu. E. and Belov, Yu. Ya. (1989). On unique solvability of an inverse problem for a parabolic equation. *Sov. Math. Dokl.,* **39,** 601–605.

Anikonov, Yu. E. and Vishnevskii, M. P. (1994). Reduction of multidimensional inverse problems and initial boundary-value problems in Hölder spaces. *Sib. Math. Journ.,* **35**(3), 495–514.

Anikonov, Yu. E. and Lorenzi, A. (1994). Existence results for parabolic identification problems. *J. Inverse Ill-Posed Probl.,* **2**(3), 181–210.

Belov, Yu. Ya. (1991). On an inverse problem for a semilinear parabolic equation. *Sov. Math. Dokl.,* **43**(1), 216–220.

Belov, Yu. Ya. and Savateev E. G. (1991). On an inverse problem for a parabolic equation with an unknown coefficient on the time derivative. *Sov. Math. Dokl.,* **43**(1), 176–180.

Belov, Yu. Ya. (1993). Inverse problems for parabolic equations. *J. Inverse Ill-Posed Probl.,* **1**(4), 283–305.

Berkhin, P. E. (1974). One boundary-value problem for equations of the mixed type. *Dokl. Akad. Nauk SSSR,* **214**(3), 496–498 (in Russian).

Beznoshchenko, N. Ya. (1975). Some problems of determination of coefficients before smallest terms of parabolic equations. *Sib. Math. J.* **16**(6), 1135–1147 (in Russian).

Beznoshchenko, N. Ya. (1979). On existence of solution of the problem of determination of the coefficient q in the equation $u_t - \Delta u + qu = F$. *Diff. Equat.,* **15**(1), 10–17 (in Russian).

Bizhanova, G. I. and Solonnikov, V. A. (1993). On solvability of initial boundary-value problem for a second-order equation with the time derivative in the boundary condition in the weight Hölder space. *Algebra and Analysis.* **5b**(1), 109–142 (in Russian).

Blokhin, A. M. (1995). *Elements of Theory of Hyperbolic Systems and Equations.* Izd. Novos. Gos. Univ., Novosibirsk (in Russian).

Bubnov, B. A. (1986). Existence and uniqueness of solutions of the inverse problems for parabolic and elliptic equations. *Nonclassical Equations of Mathematical Physics.* Izd. Sib. Otdel. Akad. Nauk SSSR. Inst. of Math., 25–29 (in Russian).

Bubnov, B. A. (1987). *On the Question of the Solvability of Multidimensional Inverse Problems for Hyperbolic Equations.*Preprint No 87–714. Computing Center, Siberian Branch of the USSR Acad. Sci., Novosibirsk (in Russian).

Cannarsa, P., Da Prato, G, and Zolezio, S. -P. (1990). The damped wave equation in a moving domain. *J. Different. Equat.,* **85**(1), 1–16.

Daynyak, V. V. and Korzyuk, V. I. (1987). The problem of the Dirichlet type for a linear differential equation of the third order. *Differential Equations,* **23**(5), 867–872 (in Russian).

Dubinskii, Yu. A. (1972). On some noncoercive nonlinear equations. *Math. Sbornik,* **87**(3), 315–323 (in Russian).

Dubinskii, Yu. A. (1973). On some differential operator equations of an arbitrary order. *Math. Sbornik,* **90**(1), 3–22 (in Russian).

Dzhuraev, T. D. (1979). *Boundary-Value Problems for Equations of Mixed and Mixed-Composite Type.* FAN, Tashkent (in Russian).

Dzhuraev, T. D. and Popelek, Ya. (1991). On classification and passing to the canonical form of equations with partial derivatives of the third order. *Differential Equations,* **27**(10), 1734–1745 (in Russian).

Engl, H. W., Scherzer O., and Yamamoto M. (1994). Uniqueness and stable determination of forcing terms in linear partial differential equations with overspecified boundary data. *Inverse Problems,* **10**(6), 1253–1276.

Fichera, G. (1960). On a unified theory of boundary-value problems for elliptic-parabolic equations of second order. *Boundary Problems of Differential Equations.* Madison Univ., Wisconsin Press. pp. 97–120.

Grisvard, P. (1967). Equations operationelles abstractes et problemes aux limites. *Ann. Sc. Norm. Super. Pisa.,* **XXI**(3), 303–347.

Hadamard, J. (1933). Proprictes d'une equation linearize aux derivees partielles du quatrine ordre. *The Tohoku Math. J.,* **37**, 133–150.

Hadamard, J. (1936). Equations aux derivees partielles. *Lenseignement Mathematique.* **35**, 5.

Kabanikhin, S. I. (1988). *Projection-Difference Methods for Determining the Coefficients of Hyperbolic Equations.* Nauka, Novosibirsk (in Russian).

Karatoprakliev, G. D. (1977). To the theory of boundary-value problems for the mixed-type equations in multidimensional domains. *Differential Equations,* **13**(1), 64–75 (in Russian).

Keldysh, M. V. (1951). On certain cases of degeneracy of equations of elliptic type on the boundary of the domain. *Dokl. Akad. Nauk SSSR,* **77**(1), 181–183 (in Russian).

Kerimov, T. M. (1977). On classical solution of a mixed problem for second-order elliptic equations. *Uspekhi Math. Nauk,* **32**(6), 255–256 (in Russian).

Kozhanov, A. I. (1992). On properties of solutions for a certain class of pseudoparabolic equations. *Dokl. Akad. Nauk SSSR,* **236**(5), 781–786 (in Russian).

Kozhanov, A. I. (1994a). Certain classes of Generate Sobolev-Gal'perin Equations. *Siberian Adv. Math.,* **4**(1), 65–94.

Kozhanov, A. I. (1994b). On boundary-value problems for certain classes of high-order equations nonsolved relative the high derivative. *Sib. Math. Journ.,* **35**(2), 359–376 (in Russian).

Kozhanov, A. I. (1997a). Existence of "almost regular" solutions of the boundary-value problem for one class of linear Sobolev equations of the odd order. *Mat. Zametki Yakutsk. Gos. Univ.,* **4**(1), 29–37 (in Russian).

Kozhanov, A. I. (1997b). Existence of regular solutions for a certain class of equations of Sobolev type of variable direction. *Mat. Zametki Yakutsk. Gos. Univ.,* **4**(2), 39–48 (in Russian).

Kozhanov, A. I. (1997c). Questions of posing and solvability of linear inverse problems for elliptic equations. *Journ. Inverse Ill-Posed Problems,* **5**(4), 337–352.

Kreiss, H.-O. (1970). Initial boundary value problem for hyperbolic system. *Comm. Pure Appl. Math.,* **23,** 277–298.

Kuczma, M. (1968). *Functional Equations in a Single Variable.* Warczawa.

Kuz'min, A. G. (1985). Some boundary-value problems for equation of the mixed type. *Differential Equations,* **21**(1), 77–83 (in Russian).

Ladyzhenskaya, O. A., Ural'tseva, N. N., and Solonnikov V. A. (1967). *Linear and Quasilinear Equations of Parabolic Type.* Nauka, Moscow (in Russian).

Ladyzhenskaya, O. A. and Ural'tseva, N. N. (1973). *Linear and Quasilinear Equations of Elliptic Type.* Nauka, Moscow (in Russian).

Lar'kin, N. A. (1980). Solvability of boundary-value problem for the stationary BT-equation. *Dynamika Sploshnoy Sredy,* **46**, 71–80 (in Russian).

Lar'kin, N. A. (1983). On the linearized equation of nonstationary gas dynamics. *Nonclassical Equations and Mixed-Type Equations.* Instit. Math., Siberian Branch of the USSR Acad. Sci., Novosibirsk, 107–118 (in Russian).

Lorenzi, A., and Prilepko, A. (1992). Fredholm-type results for integrodifferential identification of parabolic problems. *Universita Degli Studi di Milano, Quaderno,* **17**.

Mamontov, E. V. (1978). On equations of small perturbations in nonstationary near-sound gas flow. *Dynamika Sploshnoy Sredy,* **46**, 71–80 (in Russian).

Maslov, V. D. and Mosolov, P. P. (1990). *Equations of One-Dimensional Barotropic Gas.* Nauka, Moscow (in Russian).

Meredov, M. (1981). On a multidimensional equation of mixed-composite type. *Differential Equations,* **17**(6), 1122–1127 (in Russian).

Nakhushev, A. M. (1995). *Equations of Mathematical Biology.* Vysh. Shkola, Moscow (in Russian).

Oleinik, O. A. and Radkevich, E. V. (1969). *Second-Order Equations with Nonnegative Characteristic Form. Mathemmatical Analysis.* **VINITI**, Moscow (in Russian).

Oskolkov, A. P. (1975). On certain model nonstationary systems of the theory of non-Newtonian liquids. *Trudy MIAN.* **CXXVII**, 32–57 (in Russian).

Oskolkov, A. P. (1980). On the theory of Foight liquid. *Zap. Nauchn. Semin. LOMI.* **96,** 233–236 (in Russian).

Prilepko, A. I. and Orlovskii D. G. (1985). Determination of the evolution parameter of an equation and inverse problems of mathematical physics. *Differential Equations,* **21,** 96–105 (in Russian).

Prilepko, A. I. and Vasin L. A. (1989). On an inverse initial-boundary value problem for the nonlinear Navier – Stokes system in the case of final overspecification . *Diff. Uravn.,* **25,** 2164–2177 (in Russian).

Prilepko, A. I. and Kostin, A. B. (1992). On inverse problems of determination of coefficients in a parabolic equation I. *Sib. Math. Journ.,* **33**(3), 146–155 (in Russian).

Prilepko, A. I. and Kostin, A. B. (1993). On inverse problems of determination of coefficients in a parabolic equation II. *Sib. Math. Journ.,* **34**(5), 147–162 (in Russian).

Riganti, R. and Savateev, E. (1991). On the solution of an inverse problem for the nonlinear heat equation. *Rapporto Interno.,* **25**, Politecnico di Torino, Turin.

Romanov, V. G. (1984). *Inverse Problems of Mathematical Physics.* Nauka, Moscow (in Russian).

Sakamoto, R. (1970). Mixed problem for hyperbolic equation. *J. Math. Kyoto Univ.,* **I.,** **10**(3), 349–373, **II.**, 403–417.

Salakhitdinov, M. S. (1974). *Equations of the Mixed-Composite Type.* FAN., Tashkent (in Russian).

Savateev, E. G. (1993). An inverse problem for the Burgers equation and its hyperbolic regularization. *J. Inverse Ill-Posed Probl.,* **1**(3), 231–243.

Savateev, E. G. (1994). Well-posedness and reduction of an inverse problem for a hyperbolic equation. *J. Inverse Ill-Posed Probl.,* **2**(2), 165–180.

Shelukhin, V. V. (1992). *Nonlocal with respect to time problems of the hydrodynamics equations and variational principles.* Summary of Ph. D. Thesis. Novosibirsk (in Russian).

Sobolev, S. L. (1950). *Some Applications of Functional Analysis in Mathematical Physics.* Izd. Leningr. Gos. Univ., Leningrad (in Russian).

Soloviev, V. V. (1992). Solution's existence in the problem of determination of the coefficient in a half-linear equation of parabolic type. *Differential Equations,* **28**(12), 2101–2110 (in Russian).

Terekhov, A. N. (1980). *On well-posed problems of equations of mixed and mixed-composite types.* Summary of Ph. D. Thesis. Novosibirsk (in Russian).

Vishnevskii, M. P. (1992). On solvability of an inverse problem for a parabolic equation with the convergence. *Sib. Math. Journ.* **33**(3), 42–49 (in Russian).

Vragov, V. N. (1977). On the theory of boundary-value problems for equations of the mixed type in space. *Differential Equations,* **13**(6), 1098–1105 (in Russian).

Vragov, V. N. (1978). On a setting and solvability of boundary-value problems for the equations of mixed-composite type of a high order. *Mathematical Analysis and Mathematical Problems Close to Them,* Nauka, Novosibirsk, pp. 5–13 (in Russian).

Vragov, V. N. (1983). *Boundary-Value Problems for Nonclassical Equations of Mathematical Physics.* Izd. Novosib. Gos. Univ., Novosibirsk (in Russian).

Vragov, V. N. (1987). On a mixed problem for a system of hyperbolic equations. *Applic. of Funct. Analys. in Math. Phys. Inst. Math.,* Novosibirsk, pp. 59–66 (in Russian).

Vragov, V. N. (1989). On a mixed problem for the wave equation in a layer with the time derivative in the boundary conditions. *Boundary-Value Problems for Nonclassical Equations of Mathematical Physics. Inst. Math.,* Novosibirsk, pp. 100-109 (in Russian).

Bei Fragen zur Produktsicherheit wenden Sie sich bitte an:
If you have any questions regarding product safety,
please contact:

Walter de Gruyter GmbH
Genthiner Straße 13
10785 Berlin
productsafety@degruyterbrill.com